MODEL-BASED SYSTEM ARCHITECTURE

WILEY SERIES IN SYSTEMS ENGINEERING AND MANAGEMENT

Andrew P. Sage, Founding Editor

A complete list of the titles in this series appears at the end of this volume.

MODEL-BASED SYSTEM ARCHITECTURE

TIM WEILKIENS
JESKO G. LAMM
STEPHAN ROTH
MARKUS WALKER

OMG SysML, OMG Systems Modeling Language, UML, Unified Modeling Language, Model Driven Architecture, MDA, Business Motivation Model, BMM and XMI are trademarks or registered trademarks of the Object Management Group, Inc. in the United States, the European Union and other countries.

Cameo is a registered trademark of No Magic, Inc.

FAS is a registered trademark of oose Innovative Informatik eG in Germany.

SYSMOD is a registered trademark of Tim Weilkiens in Germany.

V-Modell is a registered trademark of Bundesrepublik Deutschland in Germany.

TOGAF® is a trademark of The Open Group.

Published by John Wiley & Sons, Inc., Hoboken, New Jersey Published simultaneously in Canada.

For general information on our other products and services or for technical support, please contact our Customer Care Department within the United States at (800) 762-2974, outside the United States at (317) 572-3993 or fax (317) 572-4002.

Wiley also publishes its books in a variety of electronic formats. Some content that appears in print may not be available in electronic formats. For more information about Wiley products, visit our web site at www.wiley.com.

Library of Congress Cataloging-in-Publication Data:

Weilkiens, Tim.
  Model-based system architecture / Tim Weilkiens, Jesko G. Lamm, Stephan Roth, Markus Walker.
     pages cm. – (Wiley series in systems engineering and management)
  Includes index.
  ISBN 978-1-118-89364-7 (hardback)
  1. System design. 2. Computer simulation. 3. SysML (Computer science) I. Lamm, Jesko G., 1976- II. Roth, Stephan, 1968- III. Walker, Markus, 1965- IV. Title.
  TA168.W4338 2016
  620'.0042–dc23
                                        2015024774

Typeset in 11pt/13pt TimesTenLTStd by SPi Global, Chennai, India

Printed in the United States of America

10 9 8 7 6 5 4 3 2

3   2016

Contents

Foreword

Contrary to popular myth, models are not new to systems engineering. Models are the way engineers analyze both problems and solutions, so systems models are as old as systems engineering itself. With the traditional focus on written specifications as the "source of truth," models were secondary and descriptive – sometimes reflected as simple sketches, sometimes shown in formal diagrams, partially captured in analysis packages, and often trapped in the mind of the chief engineer. The transformation of systems engineering from document-centric to model-centric practices is not about the introduction of models. It is about making models explicit and moving them to the foreground where they serve as the authoritative tool for design, analysis, communication, and system specification.

Organizations today are investing heavily in representations, standards, methodologies, and technologies to transform the practice of systems engineering through model-driven paradigms. To manage the complexity of today's problems; to keep pace with today's rapidly evolving technologies; to capture the required knowledge regarding the problem, solution, and rationale; to respond effectively to change – all require that systems engineering join the other engineering disciplines in moving beyond document-centric techniques and embracing the power of a model-based foundation. With energy and focus over the last ten years has come notable progress. The industry has advanced in the area of representations with the development of SysML as a standardized set of diagrams to complement traditional systems representations. Numerous books – including a frequently cited guide by Tim Weilkiens – explain the details of using this notation to capture and communicate system designs to improve explicitness and alignment within the systems team. Alongside these representations have emerged countless standards and frameworks to help engineering teams develop high-fidelity models reflecting key systems dimensions.

However, for all the industry discussion regarding SysML, representations, standards, and tools, there remains a great deal of confusion. Understanding SysML notation and drawing SysML diagrams do not equate to doing model-based systems engineering. The use of disjoint models and simulation in systems engineering is also not equivalent to integrated model-based systems engineering.

Effectively moving forward with the transition to model-centric techniques requires that we step back to understand the bigger picture. Diagrams and other representations do not live in isolation but are interrelated and overlapping, communicating key aspects of the system model from specific viewpoints. System architecture and detailed analytical models are not disjoint, nor is there a single grand unified model to capture all dimensions of interest for all systems problems. To move forward, we must embrace the holistic systems perspective and apply it to model-based systems engineering, seeking out the interrelationships and developing a robust toolbox of supporting practices.

In this book, Tim Weilkiens, Jesko Lamm, Stephan Roth, and Markus Walker broaden our vision and expose us to a rich set of perspectives, processes, and methods so that we can develop an effective unified framework for model-based systems architecture. Building upon the existing industry library of textbooks on SysML, this book looks beyond the representation to address models, viewpoints, and views as part of a modern approach addressing requirements, behavior, architecture, and more. It connects to a larger framework of processes, methods, and tools key to enabling model-centric practices. And it looks beyond the technical space to the critical cultural dimensions because the transformation to model-centric techniques is far less a technical challenge than one of organizational change. Addressing the broader framework, Tim, Jesko, Stephan, and Markus bring model-centric practices together to help practitioners develop cohesive system architectures – our one chance in the life of a program to manage complexity, develop resilience, and design in critical concerns such as system security.

There is no doubt that the future of systems engineering is model-based. Document-centric techniques simply are not enough as we grapple with the challenges of today and tomorrow. Those practitioners and organizations who are early adopters in developing a cohesive model-centric framework of processes, methods, and tools will certainly be at a competitive advantage – whether producing products themselves or delivering systems services for others. If, as a profession, we can transform from document-centric to model-based systems engineering and do so with the vision of enabling model-based engineering, we can help transform the larger product lifecycle delivering radical improvements in quality, cost, and time to market for the benefit of all.

David Long
President, Vitech Corporation
INCOSE President (2014 and 2015)

Preface

Reacting to market needs on time with systems of high quality and marketable costs is a strong competitive advantage. Once a market need has been identified, multiple disciplines are involved in developing a system toward it. They need to collaborate closely and each according to a precise understanding of the own contribution to the system development. Effective communication and the creation of understanding for the whole system of interest are keys for the success. Organizations are facing a more and more dynamic environment, and at the same time, an increasing organizational complexity of distributed teams and stakeholders, and an increasing technical complexity of more heterogeneous relationships between system components and their environment. This context requests an explicit and sustainable system architecture.

Each of the engineering disciplines contributing to system development needs specific views for obtaining the needed insight. System models enable the creation of consistent sets of stakeholder-specific views. People using them gain a fast and comprehensible understanding of the system they are developing, which can help them choose appropriate solutions for fulfilling the market needs. All the views look on the same data baseline. There is no effort to consolidate redundant data or to clarify misunderstandings of inconsistent information and the costs of resulted errors.

A system architect needs to shape the system architecture well for realizing a successful system. Multiple tasks have to be carried out, each using an effective approach. This book provides a toolbox for the architect for his daily challenges. The scope of the book is a model-based environment, either it is already established and running or planned. The book explains how to use the SysML modeling language in obtaining model-based architecture descriptions. Nevertheless, the concepts are independent of SysML and could also be performed with other modeling languages.

This book is about people, models, and better products, based on our belief that model-based systems architecting produces better products by creating communication and insight for people involved in system development. The book presents a collection of methods and approaches which we see as ingredients for getting the system architecture work done successfully. We present model-based system architecting, which we see as a required backbone for excellent systems architecture work together with the stakeholders. We will show that involving the stakeholders means much more than running through a formalized review process.

A fundamental principle in system architecture is simplification. Without simple concepts to be communicated to the stakeholders, the system architect will not be understood and thus will fail. We advise you, dear reader, to adopt the principle of simplification and apply it to the multitude of approaches presented in the book. Feel free to only choose the approaches that are most suitable for your daily work and disregard the others, until you are in a situation where they turn out to be useful. The book is a well-stocked toolbox and not a rigid all-or-nothing process for system architects.

Our experience tells us that each organization will have a different focus area and will need different approaches. This is why we have bundled a variety of approaches we have observed being applied successfully in the industry, in the hope that you will find some pieces of information that are suitable exactly to your current activities. We have selected those approaches that we find easy to apply in daily work and are important for common model-based system architectures. We do not claim to provide a complete set. Every system architect loves to go to a hardware store to extend her toolbox. And from time to time, she has to discard one of her tools when it is no longer appropriate.

The book addresses system architects and their managers as well as engineers who are involved or interested in systems architecting. It is the first comprehensive book that combines the emergent discipline systems architecting with model-based approaches with SysML and puts

together puzzle pieces to a complete picture. Highlighted puzzle pieces
are

- functional architectures and the Functional Architecture for
  Systems (FAS) method by Lamm and Weilkiens to derive the
  architecture from common use case analysis.
- the integration of the concept of layered architectures from the
  software discipline in the context of system architectures.
- the modeling of system variants.
- the whole picture of different architecture kinds like functional,
  logical, and product architectures and their relationships.
- a brief description of SysML and
- a summary of the history of the V-model and recent thinking about
  it in the appendix.

As a typical reader of this book, you may have no time to read all
chapters in sequential order. Therefore, we have made the chapters as
independent from each other as we could, in order to enable you to
read them individually or out of a dedicated sequence when you like
inspiration about a certain topic. You can find on-demand reference
about particular topics and get inspiration for directly using the presented approaches in your daily business. The topics are demonstrated
using a fictitious robot-based solution for virtual museum visits as an
example system.

We like to write texts using gender-fair language. On the other hand,
we avoid to clutter the flow of reading by using always both genders in
the same sentence. Therefore, we have only used one gender where it
was not appropriate to use gender-neutral language. Feel free to replace
he by she and she by he wherever it is appropriate.

We like to thank the "FAS" and "MkS" working groups of GfSE,
the German chapter of INCOSE. The work in these groups has provided us with new ideas that can now be found in this book. We thank
NoMagic for their support in working with the Cameo tool family that
we used to create the SysML models and diagrams we used in multiple
chapters of this book. We also thank Erik Solda for allowing us to use
the robot example, Martin Ruch for contributing ideas about the assessment of organizational interfaces and all the colleagues at works, who
have influenced our way of thinking, helped us with foreign languages in
both reading and writing or recommended literature that is today part of
the foundations of this book. We furthermore thank numerous people

who provided us with advice after we had shown or explained them little fragments of this book to hear a second opinion.

We like to thank all the supporters of MBSE who believe that MBSE enables the successful development of complex systems. In particular, David Long, who is a great expert of MBSE from the very beginning and has written the foreword.

Finally, we like to thank Brett Kurzman, editor at Wiley, his assistants Alex Castro and Kathleen Pagliaro, and Bhargavi Natarajan for their support.

TIM WEILKIENS,
JESKO G. LAMM,
STEPHAN ROTH,
MARKUS WALKER

Contributor
MATTHIAS DÄNZER,
Bernafon AG
February 2015

About the Companion Website

This book is accompanied by a companion website:

www.mbse-architecture.com

The website includes:

- High resolution version of all the figures in the book.

Chapter 1

Introduction

Model-based system architecture (MBSA) combines the two key technologies: model-based and system architecting. Both are major parts of the future state of systems engineering [57].

Many systems result from an evolutionary development. They are driven by their parts and do not emerge from the architecture. The parts could be anything that in combination are assembled to a man-made technical system. System architecture is exhibited by a complete system. Often system architecture is referred to the architecture from the perspective of a software architecture in combination with the hardware or the architecture of software-intensive systems [20]. We understand system architecture is more holistic and also consider systems without any software. Although systems without any software that are handled with systems engineering processes and model-based system architecture concepts like described in this book are very rare, a system architecture is always present. In today's and future systems engineering, it is crucial to apply explicit system architecting for the success of the system project [57]. Chapter 4 defines the term *system architecture* within its context.

Studies clearly show that system architecting is critical for the performance and success of the system [34]. This is particularly evident for projects that require significant architectural work or rework. Due to more and more dynamic and complex markets and environments,

Model-Based System Architecture, First Edition.
Tim Weilkiens, Jesko G. Lamm, Stephan Roth, and Markus Walker.
© 2016 John Wiley & Sons, Inc. Published 2016 by John Wiley & Sons, Inc.

the system architecture must more and more withstand the changing requirements and requests for radical changes. Chapter 3 lists the benefits of system architecting.

System architecture is about establishing solutions that are checked for feasibility by the corresponding experts, about designing interfaces that are agreed from both sides, and about ensuring that the people who should know the architecture of a system have a common understanding of it. MBSA uses models for enabling the creation of healthy communication around the architecture of the system and for ensuring that the architecture is validated from different points of view. Models are a key tool to be capable of developing complex systems on time and in a feasible quality. Chapter 5 defines the term *model* and MBSA and discusses the related terms.

Models are more than graphics. There are even models without any graphical representations. Just the graphics is not modeling, but drawing. To create a model you need semantics, which you find in a modeling language. We use the international standard Systems Modeling Language (SysML) as language for the system requirements and architecture models. Appendix A gives an overview about SysML. Although we extensively use SysML in this book, our methods and concepts are independent on SysML and could also be implemented by other modeling languages.

The system architect is the one in charge of shaping the system architecture. This is a big responsibility and a big challenge. The organizations dependent on the system should carefully select the people who are allowed to architect the system—and these people's work results will be tightly monitored by stakeholders everywhere in the organization. Chapter 19 describes how system architecting could be embedded in an organization and Chapter 10 discusses the interfaces to the stakeholders of system architecting. In particular Chapter 8 introduces the adjacent discipline requirements engineering that closely collaborates with the system architecting. The SYSMOD zigzag pattern presented in Chapter 7 shows the relationship between requirements and architecture and clearly demonstrates the need for a close collaboration. Artifacts of the model-based requirements and use case analysis are important inputs for the system architects especially to elaborate a functional architecture using the so-called FAS method.

Chapter 14 is a comprehensive presentation of the Functional Architectures for Systems (FAS) method. Functional architectures are built of functions only and are independent of the physical components that implement the functions. The functional architecture is more stable than a physical architecture that depends on the steadily changing technologies. The architecture principle to separate stable from unstable parts is covered in Chapter 7 about architecture patterns and principles.

Besides the functional architecture we define and discuss further system architecture kinds: the base architecture that fixes the preset technologies and adjusts the scope for innovation, the logical architecture that specifies the technical concepts and principles, and the product architecture that finally specifies the concrete system. All three architecture kinds are physical architectures. The layered architecture is an orthogonal aspect to these architecture kinds and is presented in Chapter 9.

Another orthogonal aspect is the modeling of variants. Variability is increasingly important. The markets are no longer satisfied by commodity products. The market requests customized products that fit to personal demands of the customers. In addition, global markets with different local environments and policies require different configurations of a system. Chapter 15 presents a model-based concept to specify different product configurations.

The architecture concepts are presented with a consistent example system. The "Virtual Museum Tour" system provides virtual visits by driving with camera-equipped robots through a real museum. The system is easy to understand and at the same time is sufficiently complex to demonstrate the system architecting concepts. The system is introduced in Chapter 2.

The system architect who thinks that his job is to make a diagram and save it on a shared network drive will most probably fail. Same for the system architects who think they are the bosses of the development staff and can instruct the other engineers. It is neither an archeological job nor a chief instructor job. System architecting is a collaborative work that requires communication and soft skills. The basics for a good communication is a common language and media to transport the information. Chapter 6 covers the artifacts of the architecture documentations. In Chapter 16, we extend our scope to system of systems and architecture frameworks.

Typically, engineers are focused on the technology challenges of their job. Nowadays, communication and more general soft skills are getting more and more important capabilities. The engineering disciplines are growing together. For instance, that could be seen by the modern discipline mechatronic. And the worldwide mankind is growing together due to the Internet, other communication, and transportation technologies.

In consequence, an engineer has an increasing number of communication relationships. She is no longer successful when she only manages her technology tasks. It is also important to collaborate well with team members, stakeholders, communities, and so on. Chapter 20 gives an introduction about soft skills for engineers.

Chapter 2

An Example: The Virtual Museum Tour System

We need an example system for the demonstration of various techniques to be presented in this book. The example of our choice is based on robotics work that one of us (Jesko G. Lamm) started doing as a leisure activity during his time at university, together with his fellow student Erik Solda who had the initial idea. During the course of this work, different robots were built. One of them is shown in Figure 2.1. It is the one that inspired the authors of this book in developing a fictitious example system to be presented in the following.

For this book, we thought of a robot according to Figure 2.2. The robot would drive through a museum. A live video stream from a camera on the robot would be broadcasted. People with access to the live video stream could thus see the exposition, even when the museum is closed.

When we had completed the first chapters of this book based on this example system, we noticed that we were not the only ones who had this idea: an organization called "The Workers" actually had this idea prior to us and invented a museum robot system that is called "After Dark", because it is intended to run during the night when the museum is closed and dark. The Workers' idea of the After Dark system won the

Model-Based System Architecture, First Edition.
Tim Weilkiens, Jesko G. Lamm, Stephan Roth, and Markus Walker.
© 2016 John Wiley & Sons, Inc. Published 2016 by John Wiley & Sons, Inc.

Figure 2.1. One of the robots that was built during the robotics work by J.G. Lamm and E. Solda. © 2007, 2014 Jesko G. Lamm, reproduced with permission.

so-called "IK Prize"—including a budget that allowed them to actually build the system and offer the first virtual tour through the gallery Tate Britain on August 13, 2014 [74].

The After Dark system came to our attention quite late during the process of writing this book. Therefore, the system we use as an example here is mainly based on our own imagination and may thus be quite different from the After Dark system, for example, regarding its architecture. Still, the After Dark system has become an important source of inspiration after Tate's press release about it [17] had been discovered by one of the authors.

In our fictitious system according to Figure 2.2, one user can log in for controlling the robot via a web application on a computer or a hand-held device, like a smart phone. Multiple users can watch the video stream from the camera via a web application. The camera can zoom and move up and down. Roughly speaking, the system-of-interest is composed

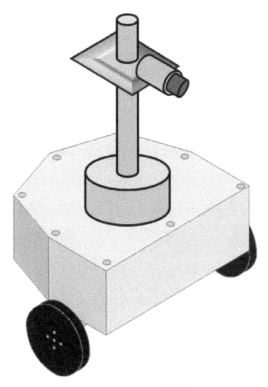

Figure 2.2. The fictitious museum robot with the purpose of enabling virtual museum tours.

of the robot itself together with the infrastructure and remote control software that is needed to operate it. We call this system the "Virtual Museum Tour system (VMT)". Chapter 9 will provide example architecture descriptions that provide a more precise definition of the system.

A little story may help understanding the capabilities of the example system: Currently, John is controlling a museum robot to drive it through a museum of arts. He has to write a report about modern art as a homework for school, and he has not had time to go to the museum during its opening hours. John types "Andy Warhol" on his smart phone and the robot starts driving to the pop arts division of the museum. Once there, it stops in the middle of a room. John now selects a painting showing a soup can. The robot moves toward the painting and stops in front of it. The camera on the robot now transmits a picture of the painting to John's smart phone. A little notification box on the smart phone displays the title of the painting. John needs to analyze the artist's way of working in more detail. Via commands entered on his smart phone, he moves the camera down. Then he zooms in on a particular area of the painting. Now he can see the necessary details via the video stream on his smart phone. This enables John to complete his homework for school.

Chapter **3**

Better Products — The Value of Systems Architecting

When driving down to the beach in a nice new car, we may enjoy how well this car grabs the road, and if mentally not yet ready for the beach we might ask ourselves which department in the car company is responsible for the driver's feeling that the car grabs the road. Is it the suspension design unit or the department with all the steering experts? We believe that all these departments alone cannot make us feel that the car grabs the road, because to do so, the car manufacturer has to see the "car as a system, as a collection of things that interact with each other and with the driver and the road" [93]. Systems architecting will ensure that the interactions between components are controlled in a proper way and that components are designed to fit each other.

3.1 THE SHARE OF SYSTEMS ARCHITECTING IN MAKING BETTER PRODUCTS

The example of the car that grabs the road was given by J.N. Martin [93]. We extend this example by speculating what would happen if different developers of car components were asked on whose merit is that the

Model-Based System Architecture, First Edition.
Tim Weilkiens, Jesko G. Lamm, Stephan Roth, and Markus Walker.
© 2016 John Wiley & Sons, Inc. Published 2016 by John Wiley & Sons, Inc.

car grabs the road. As a reply, maybe the car manufacturer's suspension department and steering experts as well as the tire companies would claim that they are the ones, by making the best possible suspension, steering, tires, and so on, but in contrast to this, consider the following example by Russell L. Ackoff: "Suppose we bring one of each of these [many existing types of automobiles] into a large garage and then employ a number of outstanding automotive engineers to determine which one has the best carburetor. When they have done so, we record the result and ask them to do the same for engines. We continue this process until we have covered all the parts required for an automobile. Then we ask the engineers to remove and assemble these parts. Would we obtain the best possible automobile? Of course not." ([2], p. 18).

Since systems architecting is concerned with making components fit together instead of making them "the best" each on its own, we believe that systems architecting is an approach that will help an organization think, develop, produce, and maintain better products.

3.2   THE BENEFITS THAT CAN BE ACHIEVED

When talking of better products, then "better" can have two different meanings:

- More satisfying or even more enjoyable for the customer (as shown with the "grabs the road" example).
- More profitable for the organization.

Of course the first aspects may lead to the second one because products that are well received by the customer have the potential to become best-sellers and thus to generate profit for the organization.

It cannot be stated in general whether an organization sees it as most beneficial to minimize development cost, production cost or maximize user satisfaction or certain quality measures. In the end, trade studies will determine the optimum trade-offs between these different criteria and probably many more. Systems architecting is the activity that will produce well-founded trade studies, because it is right in between the requirements analysis work and the solution space that is framed by the development of different system elements. Figure 3.1 illustrates this based on the example of system-level trade studies across subsystems. Systems architecting can enable a top-down realization of business goals, requirements, quality criteria, and product strategies into

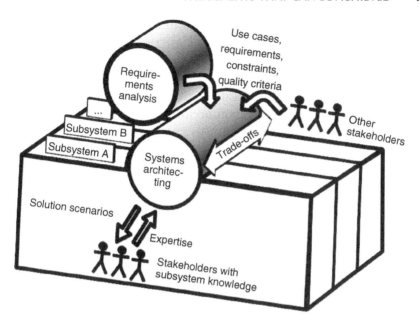

Figure 3.1. Systems architecting plays a major role in breaking down requirements to solutions and in optimizing solutions for example by finding good trade-offs.

solutions. This is almost independent from whether a completely new system is developed (a very rare case) or whether an existing system needs to evolve further, by architecting increments to an existing solution. The only difference is that constraints from the existing solution will enter trade studies in systems architecting in the latter case, via the expertise from subsystem development.

It is via this top-down approach that usability, maintainability, reliability, and so on can be designed into the system and that concepts like design for user experience [51] or design to market [142] can be realized.

3.2.1 Benefit for the Customer

Different kinds of businesses have different customers: while the consumer goods industry targets millions of individual consumers, subcontractors target industries whose suppliers they are. Despite the variety of customers addressed by different industries, we believe that any system developer will increase customer satisfaction with investments into systems architecting.

The good news for industrial customers of subcontractors is as follows: we expect change requests and risk management to work very smoothly on a well-architected system with a proper architecture description in place. For example, we have seen cases in which the impact of a change could be easily analyzed, based on the architecture description, and since the system architecture captures dependencies inside the system we also expect it to help analyzing how uncertainty in one area of the system may lead to risks in other areas. Being able to manage risk based on the knowledge of the system architecture is indeed a potential sales driver, if we believe in one of the conclusions of J.P. Monat's article "Why Customers Buy" [97]: customers' perception of risk seems to be an important factor in the purchasing decision.

The good news for the consumer of mass products is as follows: we expect a properly architected system to have a good chance of working as expected in the market place, because systems architecting can allow for thinking the different modes of operation into it instead of just testing them into it, and well-defined interface descriptions may offer a basis for planning the systematic review or testing of component interactions in order to find flaws. This is particularly interesting with regard to discovering rare cases for which interfaces are not prepared. Well-documented architecture also enables continuous reassessment of quality, for example, by means of reviews and hence is a good basis for preventing flaws introduced by continuous changes during product improvements. This in turn may lower the threshold for continuous changes with the purpose of continuous product improvement.

Last but not least, the properly architected system has a good chance of achieving an attractive cost-benefit ratio, because the organization that developed the system and the production facility producing it could save development and production cost, as it will be explained in the next section.

3.2.2   Benefit for the Organization

Each system exhibits an architecture (ISO/IEC/IEEE 42010:2011 [64]) or in other words, every system development produces system architecture. The question is whether the system architecture evolves implicitly or whether it is explicitly defined.

The precondition for the benefits to be discussed is an explicit way of system architecting, where system development explicitly involves system architecture processes. In case these are not yet established, there will be an initial investment into architecture work, before the organization can harvest the expected benefits.

Once an organization has established systems architecting as an integral part of system development, it should see the predictability and efficiency of system development increase and it should see cost decrease.

Predictability should be obtained because system architecture supports planning and risk management:

- Planning is supported, because the knowledge about the system's architecture enables a completeness check of the work breakdown for the development work and the identification of dependencies between work packages. It is also supported because the order of integration and the needed verification can be planned and optimized according to knowledge about the system architecture.
- Risk management is supported for example because the system architecture determines the contribution of subsystems to the system's performance and thus needs to be known for quantifying the influence of risks in subsystem development on the overall risk profile of system development.

Efficiency should be gained, because systems architecting can make system development shorter and even lead to reduced production cost. Shorter development time should be achieved because systems architecting can ensure that subsystem development is based on mature interface specifications, which reduces the risk of failures during system integration. Production cost can be reduced if production is thought into system development right from the start. Only with a system-level approach, one can effectively avoid errors in the assembly line, such as inefficient order of assembly steps. For example, there is a constraint on the order of steps in production, if the system is designed to have all the programming interfaces for production inside the housing, where they are inaccessible once the system is fully assembled. If production engineers can save cost by using the programming interface after the last assembly step, then a redesign of such a system may be triggered. Systems architecting should ensure that manufacturability is thought into the system in early phases of the life-cycle, such that a redesign as described is either not even necessary or happens "on paper" during early concept phases, thus before any money has been invested into engineering the components under redesign.

Cost should decrease, because systems architecting allows to steer system development into a suitable direction right from the start, leading to a focused investment into development efforts. Thinking of Ackhoff's car example from the beginning of this chapter again, one might compute how expensive it can be to build the best possible variant of each part of the car. A very expensive endeavor, particularly because Ackoff's example tells us that the resulting car will be quite poor, meaning that even more cost will arise from making the car as a whole better. Systems Architecting can help reducing cost by ensuring that subsystems are developed and produced

- according to the correct specification,
- in just the quality that is needed, and not in worse or (even more expensive) better quality.

We also expect systems architecting to reduce cost, because it facilitates communication in and between development teams, spreads knowledge, makes knowledge accessible, and thus ensures that coordination effort stays low. As a result, an investment into system architecting may lead to reduced cost and thus to return of investment, but also to shortened time-to-market (which can again contribute to return-of-investment, e.g., if the battle against competitors is won in terms of being the first on the market with a certain asset). Of course an investment into system architecting is in the first place a driver of cost for organizations not practicing systems architecting yet. New ways of working have to be established via a costly change process. To be able to harvest, it may thus be necessary to accept higher cost in early phases, based on belief in the approach.

Finally, it may simply be more satisfactory for the engineers in an organization to work based on a well-defined systems architecture and spend their brainpower into making it right in the first place than to run after all interactions inside a system, while they are being discovered on the fly. If done right, systems architecting can be fun for all involved stakeholders.

3.3   THE BENEFITS THAT CAN BE COMMUNICATED INSIDE THE ORGANIZATION

In the struggle for budget and resources, different entities in the organization try to create transparency about their contribution to business success. System architects have a particular challenge in doing so, because their contribution is usually indirect. Architecture descriptions cannot be sold to the customer, unless the organization sells systems engineering consulting. System architects will thus have to convince the organization of their important contribution to making better products.

Earlier, it was mentioned that the suspension department in a car company will claim having a major share in how well the sold cars grab the road. The suspension experts in such a department may be insulted if the organization is told that this was the system architects' merit. Simply saying "System Architects are the ones that care about better products" therefore is probably not the right statement to make.

In our experience, it is the benefits for the organization that will in the end convince others of the value of systems architecting. In this case, it is certainly not sufficient to write these on a big poster or present strategy slides that list these benefits. The positive effect of systems architecting has to be "felt" in different parts of the organization. The best feedback a system architect can get is a testimonial from a key player in engineering, stating for example how well a certain interface definition activity has made different fields of engineering reach a common understanding and a well-functioning interface agreement.

In our work in systems architecting, we have also experienced that it is the unintended positive side effects of systems architecting that lead to affirmative testimonials from within the organization. How often have we seen the following? We called in several development stakeholders with the intention of making a very small detailed discussion. Then, during the course of action, it was revealed that there was a fundamental misunderstanding between two or more parties within the development organization. More than once have some of the authors received feedback as follows: "this was good, hadn't you called us in then we would have continued our work based on the wrong assumption."

We recommend to record such feedback and to spread affirmative testimonials, because we believe that the strong benefits for the customer and the organization can only be created in an organization where the benefits of systems architecting are well recognized and where different parts of the organization are supporting systems architecting.

3.4 THE BENEFICIAL ELEMENTS OF SYSTEMS ARCHITECTING

The system architect should be aware which deliverables from systems architecting provide benefit to the organization and its customers. There are some deliverables that do not provide the benefit on their own.

For example, the system architecture per se does not necessarily provide customer benefits. It is something that is exhibited by the system and is not explicitly requested by the customer—at least not to a full extent. But the concepts realized in the system architecture can result in customer satisfaction or reducing time to market. This can provide benefit for customers or organizations, respectively. The architecture description again does not provide benefits by itself, since it is residing in some repositories and just consuming memory. Only by using the architecture description to increase clarity about the development task at hand can the system architect contribute to avoiding wrong development approaches and flawed products. The system architect thus has to communicate with the system architecture stakeholders to make the organization harvest the benefits of having created a system architecture description.

So in summary, it is what we do around the system architecture and its description that creates the benefit. In turn this means that before investing any time into a certain systems architecting task, the system architect should assess what to do with the work product from the task. Only if the work product of a systems architecting task can be used for beneficial activities and only if these activities fit into the schedules of the organization should the given task be started.

3.5   BENEFITS OF MODEL-BASED SYSTEMS ARCHITECTING

The system of interest exhibits a system architecture (ISO/IEC/IEEE 42010:2011 [64]). The aim of systems architecting is to shape the architecture that is later exhibited by a system under development. This cannot be accomplished by the architect alone but only by means of a close cooperation between architects and architecture stakeholders, in particular engineering domains and their representatives and developers. If those develop the system according to an agreed architecture description, then the described and exhibited system architecture are aligned.

To ensure that architecture stakeholder have a consistent understanding of the architecture to be exhibited by a system, the architecture description has to be communicated to the architecture stakeholders by means of the appropriate architecture views. Model-based systems architecting allows creating a model from which different views can be generated for addressing concerns of different stakeholders. The model in the background ensures that the different views stay consistent and can be regenerated after major changes.

In a model-based environment, the architect can focus on finding the right views and the suitable visualizations for the work with architecture stakeholders, instead of having to be concerned about keeping different views consistent. The model becomes the single source of truth.

Once a model has been created or updated, new views can be generated on demand. For example, if the system structure is modeled, it is possible to create a list of all system elements from the model at any time and to filter it for certain aspects. One could for example create a list of all the user interfaces of the system in order to support the systematic planning of usability testing.

Another aspect of model-based systems architecting is the possibility to validate ideas by means of a model. C. Alexander writes that the way to improve a picture in the mind of a designer is to make an even more abstract picture ([4], p. 77–78). In terms of model-based system architecture, we suggest that the model takes the role of this more abstract picture as proposed by Alexander. It depends on the task at hand whether the best application of this thought is the creation of executable models that are used for simulations or whether it is the rigorous (human or automated) review of models according to well-defined criteria.

Chapter **4**

Definition of System Architecture

Defining "architecture" appears to be rather difficult considering the big number of existing definitions. They do share some commonality but in detail differences become obvious. Interestingly, definitions from different domains are debated controversially. For instance, some enterprise architects refuse to accept the definition of ISO/IEC/IEEE 42010: 2011 [64] with the rationale that an enterprise is no software intensive system. Software intensive systems are in the scope of the standard IEEE Std 1471-2000 [65], a predecessor of ISO/IEC/IEEE 42010:2011, from which the refused definition originates. Nevertheless, the definition for "architecture" in the TOGAF® version 9.1 [136] is very close to that of ISO/IEC/IEEE 42010:2011. Searching the online browsing platform of ISO [58] for the exact term "architecture" in the area "Terms & Definitions" results in 30+ hits. Considering terms comprising the word "architecture", such as "logical architecture", the search results in 120+ hits. The Software Engineering Institute (SEI) of the Carnegie Mellon University maintains another long list of definitions [128]. Though the latter lists definitions of "software architecture," it may serve as a reference when defining "system architecture."

Model-Based System Architecture, First Edition.
Tim Weilkiens, Jesko G. Lamm, Stephan Roth, and Markus Walker.
© 2016 John Wiley & Sons, Inc. Published 2016 by John Wiley & Sons, Inc.

This book can hardly provide a global accepted definition for "architecture". Just adding another definition of architecture will not add benefit to the systems engineering community. As this book promotes architecting based on models, the hereafter provided definition shall be grounded in a model as well. This follows the approach to model sentences to get formal definitions as proposed in "Architecture and Principles of Systems Engineering" by Dickerson and Mavris [29]. The proposed approach shall be extended to a set of definitions sharing the same elements to get integrated definitions for the terms "system architecture", "system" and "architecture description".

4.1   WHAT IS ARCHITECTURE? – DISCUSSION OF SOME EXISTING DEFINITIONS

What makes a Gothic cathedral what it is? What is its architecture? What is the architecture of an elevator or a hearing aid system? What makes it usable, durable, and beautiful? The last question is very old. The Roman architect Vitruvius Pollio (first century BC) wrote in his first book about architecture that buildings should be made to be durable, usable, and beautiful ("ratio firmitatis, utilitatis, venustatis" [119]).

The term "architecture" origins from "architect," the role that realizes the architecture of something. According to the Oxford Dictionary [112] derives the term "architect" from the Greek *arkhitektōn*, from *arkhi-* "chief" + *tektōn* "builder".

If architecture is a property of a system, as suggested by a number of definitions, we would need to distinguish between the description of the system and the description of the architecture. And properties of the architecture would need to be distinguished from properties of the system, though they may overlap. For instance, the structure of a system is a property of both the system and the architecture. But the architecture should drive this structure by applying principles such as organization principles to achieve intended characteristics of the system. Therefore, the applied principles become part of that architecture. This imposes that definitions relating architecture only to the structure of the system would relate to only a part of the architecture. Similar with behavior, architecture makes use of interaction principles to drive the interactions within the system and with its environment to shape behavior as intended. Consequently, if we distinguish between "system", "architecture" and "architecture description" we need to distinguish between "system description" and "architecture description" as well. Following above argumentation, the description of the system depicts the system as it is, where the architecture description adds explanation why and how the system came to its shape.

The process of architecting leads from requirements to one or more designs that satisfy the requirements. The design of a system comprises system elements. They can be obtained through development and production or purchasing (designated as implementation process in ISO/IEC 15288:2008 [60]). Integrating the system elements in accordance to the system's design leads to the built system. This sequence imposes that the terms "architecture" and "design" are synonymous. Though, the distinction between the two terms is controversially debated. Consulting a number of glossaries does indeed support the opinion that these two terms are synonyms. The term "architecture" is frequently related to high-level design, typically involving multiple disciplines to realize the system elements. The term "design" seems to be used when only a single discipline is involved to realize a system. For the modeling of definitions hereafter we rate the two terms "architecture" and "design" as synonymous.

If the process of architecting leads from requirements to an architecture or a design, it forms a bridge between the problem space described with requirements statements to the solution space in which the built system resides. Some definitions attribute architecture as necessary and sufficient to satisfy the requirements. A design comprising only necessary and sufficient elements would be the ideal solution of the system of interest. Since we do not live in an ideal world, constraints such as time, cost, and others refrain us from reaching ideal solutions. As the architecture has to consider constraints it will inherit nonideal properties. Hence architecture needs to comprise more than necessary and sufficient parts.

The architecture starts to exist as soon as we start analyzing and validating stakeholder requirements and transforming them into system requirements. We follow here the terminology of ISO/IEC/IEEE 29148:2011 [63] that makes a distinction between stakeholder requirements and system requirements. The first capture the needs of stakeholders as solution neutral as possible. The latter follow a design hypothesis and transform the stakeholder centric definitions into technical characteristics of the design hypothesis.

Initial decisions are often very implicit and related to the business model of the respective company. For instance, a stakeholder requirement may request to get inside a building from one floor to another. This could be satisfied by designing stairs. If the respective company wants to sell elevators, the imposed design is rather an elevator.

A considerable number of definitions relates "architecture" to principles, for example, [48]. Principles should drive the structure and behavior and hence are the important pillars of architecture. Principles are accepted or professed rules guiding the creation of the architecture. For instance, a Gothic cathedral follows the principle that observers follow with their eyes the lines imposed with the design. This leads the observer to watch up toward the heaven. This behavior of the observer was maybe a requirement of a stakeholder at that time to make the observer feel that the church provides the relation between earth and heaven. Changing principles late during the development will most probably result in big effort and costs. Relating architecture with principles confirms the qualitative description that architecture includes everything that is difficult or expensive to change later in the development.

Emes et al. conclude in "Interpreting 'Systems Architecting'" [36], that a single-sentence expression of a root definition is hardly sufficient to cover the diverse viewpoints relating to many domains when defining "architect". This conclusion holds true for the definition of "system architecture" as well. Modeling the definition of system architecture together with the related definitions ensures consistency and may enable understanding between involved domains.

4.2 MODELING THE DEFINITIONS OF "SYSTEM" AND "SYSTEM ARCHITECTURE"

The definitions of "system" and "system architecture" have a tight relationship. Therefore, they should be modeled together. The definitions provided by Dickerson and Mavris [29] for the terms will serve as staring point.

> "A system is a combination of interacting elements organized to realize properties, behaviors, and capabilities that achieve one or more stated purposes."

> "System Architecture is the organization of the system components, their relations to each other, and to the environment, and the principles guiding its design and evolution."

The resulting definitions are depicted in Figures 4.1 and 4.2.

To merge these definition into one model, the following terms or phrases need to be reconsidered:

- "elements" or "components": Elements that constitute a system shall be designated with the term "system element" according to

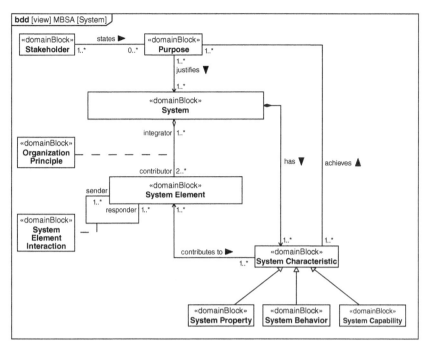

Figure 4.1. Definition of "System".

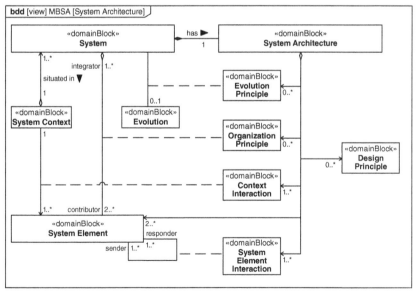

Figure 4.2. Definition of "System Architecture."

the definition in ISO/IEC 15288:2008 [60]. A "system element" may be a "system" in its own right. This relation is elided in the diagram hereafter.

- "organization" or "organized": The organization of a system should follow some principles to justify the emerging organization or structure of the system. "Organization principles" govern the relation between the system and its system elements.
- "interacting" or "relation to each other": The "system element interaction" captures how system elements operate together to achieve system characteristics.
- "relations ... , and to the environment": The system is situated in an environment. Those parts of the environment relevant for the system are designated as "system context". This imposes that some parts of the environment have negligible impact to the system. Some system element will be exposed to that system context and interact with it. The "context interaction" describes each stimulation or impact to the system from outside the system border, and vice versa, including interaction with users not considered as part of the system.
- "principles guiding its design": The "design principle" captures rules and best practices applied in the definition for system element interaction and context interaction (i.e., interaction principles) and definitions for system elements. Since we rate "architecture" and "design" as synonymous, a "design principle" could also be designated with "architecture principle".

Dickerson and Mavris represent above definitions in UML class diagrams. We use hereafter SysML block definition diagrams (Figures 4.1 and 4.2) and adapt some modeling done by Dickerson and Mavris to merge the two definitions. To make a clear distinction to models about products, we use the stereotype «domainBlock» of the SYSMOD profile for involved entities.

4.2.1 Definition of "System"

Instead of relating the term "system" to the term "combination" as in [29], we use an aggregation relationship with an association block to describe "combination of interacting elements, organized to ... ". The association block defines the organization principles according to which the system elements form the system. The aggregation association emphasizes the whole-part relation between the system and its system elements. We could now discuss if this should rather be a composition association. Indeed, for some systems a composition association would be more appropriate. For system elements with existential dependency on the system, above notation is a bit imprecise, but not wrong. But with a lot of systems the system elements are not existentially dependent on the respective systems. As discussed in "A Journey Through the Systems Landscape" [86], some systems exist only in certain views.

Regardless of the existential dependency of system elements on the system, we recommend to use a composition association for modeling a product structure. This proved to be better understandable by the stakeholders.

The role of a system element is to contribute to the system and the system integrates its system elements. Though this is most obvious, debating issues in the daily work such roles need to be emphasized now and then to reach valuable solutions.

The multiplicity of system elements has a lower bound of two. A system needs to comprise of more than one system element. Otherwise, the distinction between system and system element would not make any sense. The upper bound is any number. To facilitate understanding of the system, the architect should consider an upper bound of 7 ± 2. This relates to the capability of humans in processing data [96]. If an architect designs the system based on this rule, he/she may face fewer difficulties in explaining the system to the stakeholders. Since this is no hard rule, we intentionally used a should-statement. The architect may want to use different numbers for different kinds of systems. For structuring the system model, a higher number will be more appropriate than depicting visible parts of the system for nontechnically trained persons.

Since a system comprises interacting system elements to realize the system's properties, behavior, and capabilities, each system element will have at least one relation to another system element. The system element interaction defines how the involved system elements shall interact. These interactions may impact detail level of the system and hence may need to be updated once these detail levels evolve.

Each system element contributes to one or more characteristics (properties, behavior, or capabilities) of the system. System characteristics achieve one or more purposes. One or more purposes justify the existence of the system. A purpose may justify more than one, sometimes competing systems. The purpose of a system should be stated by stakeholders. Sometimes, stakeholder may state requirements but omit the related purpose. For proper satisfaction of stakeholders, the purpose should be known as precise as possible.

4.2.2 Definition of "System Architecture"

To express the strong relation and dependency between the system and its system architecture, a composition association describes this relation. Each system has exactly one system architecture and each system architecture belongs to one system. This definition applies to each abstraction level of a system. That is, if the system of interest is the functional system, its system architecture is the functional architecture. Section 14.7 provides definitions and relationship information on system architectures of different abstraction levels.

The system architecture may depend on a number of principles regarding its organization, the design and the system's evolution. The lower bound of multiplicity of each principle is zero to indicate that development can occur with no predefined principles. The upper bound of multiplicity is not further limited. A big number will not necessarily result in better architecture but increase the difficulty to find a valid solution. Principles can be rated as learning from earlier projects. Eliding principles increases the risk of additional effort for redesign in later life cycle stages of the system. Most systems will have an evolution. Considering the future will assist an architect when selecting appropriate principles or arguing against imposed principles.

The interactions within the system (system element interaction) and with its context (context interaction) form parts of the architecture. At least one interaction between different system elements exists. And at least one interaction between a system element and the system context exists. System elements need to interact to create the emergence of the system. And at least a technical system requires a stimulus from its context to function.

A system is situated in the system context. The system context may be dynamic and it may be different for different instances of the system. Depending on the organization of system elements, not each system element will be exposed to the system context. At least one, the outermost will be exposed.

Chapter 5

Model-Based System Architecture

The discipline model-based systems engineering (MBSE) has become very famous in systems engineering in the last years. INCOSE defines MBSE as "the formalized application of modeling to support system requirements, design, analysis, verification, and validation activities beginning in the conceptual design phase and continuing throughout development and later life cycle phases" [55]. The most recent INCOSE Vision 2025 [57] mentions system architecture as a crucial discipline for future successful systems engineering. Model-based system architecture combines both enablers.

Although the term *model* is essential, there is no common definition for the term *model* in the context of MBSE. Stachowiak defines in his book about general model theory three features of a model [130]:

- Mapping—A model is a mapping of something else.
- Reduction—A model only reflects parts of the original thing.
- Pragmatic—A model fulfills a specific function and is used in place of the original for this purpose.

Model-Based System Architecture, First Edition.
Tim Weilkiens, Jesko G. Lamm, Stephan Roth, and Markus Walker.
© 2016 John Wiley & Sons, Inc. Published 2016 by John Wiley & Sons, Inc.

The mapping and reduction features imply another often-mentioned feature of models: abstraction. *Abstraction* is the process to reduce the information about a concept to the relevant parts for a particular purpose. An abstraction is the result of an abstraction process. See also Section 11.2.5 about the abstraction skill of system architects.

We second the features of Stachowiak and add some more to give a definition how we understand and use the term *model* respectively the *system model* in the context of MBSE:

The *system model* (in the context of MBSE) is an abstraction of a real or to be realized system. The system model is characterized by the following properties:

- The entire system model may be composed out of multiple repositories, but from a user's point of view it must behave like a single, consistent model.
- The abstract syntax of the model language covers engineering concepts such as requirements, behavior, blocks, and tests.
- The model enables different kinds of views.

The definition differentiates between the model and the repository. A *repository* is a data storage like a file or database, that is, an implementation of a model. For example, a Systems Modeling Language (SysML) model could be stored in a file and a requirements model in a database. These repositories are connected for instance by using the ReqIF data exchange format [107]. The connection is discontinuous and does not exist all the time. Therefore, it is important to track the versions of the repositories to define a valid configuration. This is part of the configuration management.

From the user's point of view both repositories—the requirements database and the SysML model in a file—act like a single model. The user could navigate from the requirements to the architecture elements that satisfy those requirements and vice versa without recognizing that she crosses the border of the models (Figure 5.1).

We know that this is a challenging feature and nowadays typically not implemented in current modeling tool landscapes. However, it is covered by many research projects like the FAS4M[1] project that closes

---

[1] http://www.fas4m.de

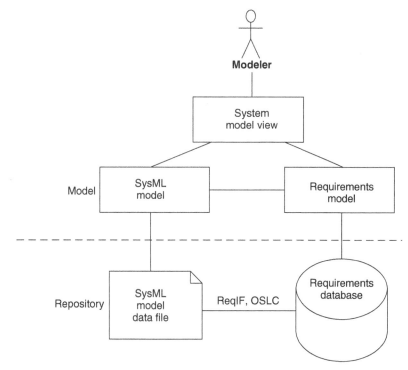

Figure 5.1. System model.

the gap between system and CAD models or the CRYSTAL project[2]. That the user does not recognize that he/she crosses the border of repositories is not a mandatory feature of a system model. It is sufficient that she can cross the borders of repositories (Figure 5.2).

The abstract syntax defines the model elements and the structure of the modeling language. It is not the same as the notation which is called concrete syntax (Section A.1). Our definition of a system model requests that the abstract syntax of the model covers engineering concepts. For instance, the abstract syntax of SysML includes elements for requirements and system blocks, while the abstract syntax of a text document covers concepts like header or paragraph. For instance, in an HTML document: `<h1>This is a header</h1><p>This is a paragraph!</p>`. Therefore, a system description in a text document is not a system model although it could be a view on a system model.

---

[2] http://www.crystal-artemis.eu/

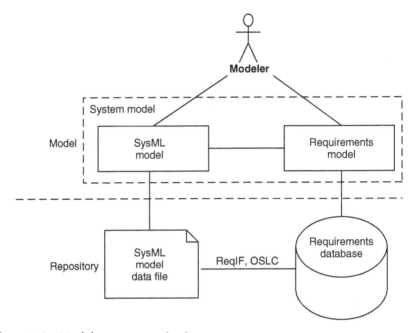

Figure 5.2. Model versus repositories.

The system model must provide different views for the different stakeholder concerns. Typically, at least one view is a text document. Other views could be for instance a set of SysML diagrams, spreadsheets, or slide presentations. It is a strength of modeling that models provide multiple views of the same consistent information to enable the communication between heterogeneous stakeholders.

A system model could be more than a SysML model. It is even not necessary to use SysML for a system model. Although we recommend SysML as the modeling language for holistic system requirements and architecture models.

Model-Based in the term MBSA means that the key artifacts of the system architecture are stored in a system model. In a document-based approach, the key artifacts are stored in documents and models are only used to contribute information to the documents, for instance, SysML diagrams in a text document.

Also known is the term *model-driven*. There exists no common definition of the term and there are different meanings around about the difference of *model-based* versus *model-driven*. Some say that *model-based* is a softer version of *model-driven*, that is, the model is

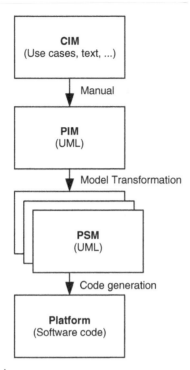

Figure 5.3. MDA levels.

an important asset, but not the source. Others say that both terms are synonyms. We do not use the term *model-driven* in this book except for a short explanation of the next term MDA.

Another term often mentioned in the context of modeling is the OMG Model Driven Architecture® (MDA®) [103]. MDA is a concept of the software engineering discipline where the model is the central asset of the development process. In a nutshell MDA defines three model levels: CIM, PIM, and PSM (Figure 5.3). All three levels are a description of the complete system, but on different levels of abstraction.

CIM stands for "Computational Independent Model". It focusses on the domain and the language of the stakeholders. Typical artifacts of a CIM are use cases and texts. PIM stands for "Platform Independent Model". A PIM is already technology oriented and more formal than the CIM, but still independent of a specific software engineering platform like Microsoft .NET® or Java2 Enterprise Edition (J2EE®). PSM stands for "Platform Specific Model". It combines the specification of the PIM with platform-specific information. A PSM could be the result of an automatic model transformation from the PIM.

The basic notion of MDA is also interesting for MBSA. The concept of different model levels is partly reflected by the separation of the logical and product architecture which could roughly be compared with a PIM respectively PSM. See Section 14.7 for a definition of logical and product architectures.

According to the definition of MBSE by INCOSE given above we define model-based system architecture:

Model-based system architecture (MBSA) is the formal application of modeling to support system architecture activities.

You can apply modeling at different levels of intensity. As an example, Figure 5.4 shows the SYSMOD intensity model [147, 145]. The three main levels define each a primary modeling goal.

Communication — The basic end of the modeling effort is to enable and improve the communication between the stakeholder of the project including the development team. On that level, you have a strong focus on the view, that is, the diagrams of the model. The model data itself is less important.

Traceability — The primary goal of modeling is the traceability between engineering artifacts. For instance, the connection from a

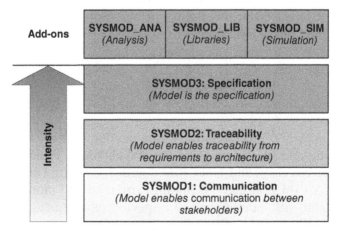

Figure 5.4. SYSMOD intensity model.

physical block of the system to the set of requirements that are satisfied by the block. It is important to have a well-structured model to achieve valuable traceability.

Specification — This level is the real MBSE. The model is the master of the key artifacts of the system requirements and architecture.

Chapter **6**

Architecture Description

6.1 WHY SPENDING EFFORT TO DESCRIBE THE ARCHITECTURE?

The only purpose of an architecture description is to explain the architecture of a system to its stakeholders. An architecture description shall document how a certain design satisfies stakeholder requirements or even stakeholder needs. The architecture description supports an architect in communication with stakeholders about the system of interest. It documents both, rationales and architecture decisions. Both may result from the architecture assessments (see Chapter 18). Therefore, an architecture description needs at least to address stakeholders that are relevant for the success of the system.

The architecture starts to exist with the initial design hypothesis while analyzing and validating stakeholder requirements. As mentioned in Chapter 4, the base architecture may be imposed by the

Model-Based System Architecture, First Edition.
Tim Weilkiens, Jesko G. Lamm, Stephan Roth, and Markus Walker.
© 2016 John Wiley & Sons, Inc. Published 2016 by John Wiley & Sons, Inc.

concept of operation (ConOps)[1] of the company, describing how the company intends to make money. Such imposed architecture is often not explicitly described. Often such architecture appears to the involved parties as obvious as it is not worth to spend effort for a more explicit description of the architecture or decisions that lead to that design. This is not necessarily wrong and may be economically justified. Though, with no tangible description of the architecture, the related rationales and principles that lead to the solution are probably not clear to each relevant stakeholder. The absence of an explicit architecture description may result in rating the current architecture as unchangeable. This imposes the risk to miss more innovative or more economic solutions. This holds true not only for the base architecture but for architecture in general.

Since the architecture description shall support the communication, it needs to be written in a language understood and accepted by each participant of that communication. The acceptance of an architecture description language by the respective stakeholders should not be underestimated. If a stakeholder does not like the form of a representation she is likely not in favor of the content of that representation.

In an ideal world, each stakeholder would be involved in this communication. The real world is mainly economically driven, so the architect will weight stakeholders in relation of their impact to the system's success. Consequently, in a lot of cases the architecture description will not be complete.

Each stakeholder has at least one concern related to the system of interest. Such concerns are grounded in the impact by the system of interest on the process operated by that stakeholder. For instance, a firefighter's process includes rescuing people and extinguishing fires. In buildings, elevators impact this process for the good and the bad. Ordinary elevators may cause additional threads by entrapped people due to power outage caused by the fire or firefighting actions. But a firefighter's elevator provides transportation capacity to improve the firefighting process. Hence, firefighters have an interest in elevators and

---

[1] We follow here the terminology of ISO/IEC/IEEE 29148:2011 [63] that makes a distinction between ConOps and OpsCon. The first refers to the intention on how to operate a business or a company whereas the latter refers to the intention on how to operate the system of interest. Other literature does not make this distinction. For example, the INCOSE Systems Engineering Handbook [56] does explicitly refrain from such distinction.

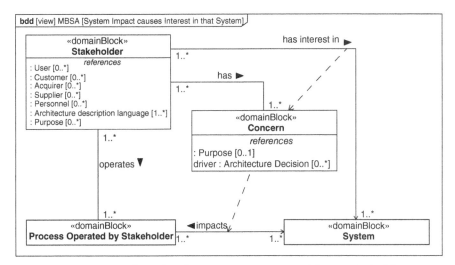

Figure 6.1. Impacts of any kind on a process operated by a stakeholder causes concerns. A concern results in an interest in the system.

become stakeholders, though they are certainly not the primary users or customers.

The processes operated by stakeholders include also the system life cycle processes. Apart from the obvious stakeholder classes such as users and customers, an architect should also consider personnel of the organization developing the system, suppliers, and acquirers. This means, each system, even if rather simple, has a considerable number of stakeholders. An architect will have to analyze the impact of these stakeholders to the system's success and decide in what quality and quantity the architecture description shall serve them. Figure 6.1 illustrates how somebody becomes a stakeholder and hence a potential consumer of the architecture description.

6.2   THE ARCHITECTURE DESCRIPTION

The definitions and explanations in this section build up on the concept of "architecture description" and related terms described in ISO/IEC/IEEE 42010:2011 [64]. They are related to the definitions on "System" and "System Architecture" provided in Chapter Definition of System Architecture.

An architecture description shall assist in explaining how concerns have been considered. That is, the architecture description shall visualize how the system of interest provides benefit or how threads by the system of interest to the stakeholder's processes are minimized. Or it shall explain why such threads could not be eliminated. For this purpose, the architecture description should comprise of four architecture description elements (AD elements):

- architecture viewpoints
- architecture views
- architecture decisions
- architecture rationales

Ideally, an architecture description includes each of these four elements. Economic reasoning may result in eliding some of them. In minimum, the architecture description shall identify

- the system
- the system's context
- the stakeholder(s) it addresses
- the concern(s) it covers
- the system element(s) it considers

Figure 6.2 depicts the definition for "Architecture Description". This definition imposes that an architecture description does not need to exist at all, though this is hardly desirable. And it imposes that more than one architecture description for a system may exist. Both can be observed in practice. The latter suggests some measures to keep different architecture descriptions in synch. Maintaining data of the system's architecture in a model eases the generation of architecture descriptions as needed and enables to keep them synchronized with only limited additional effort.

6.2.1 Architecture Viewpoint

Each stakeholder perceives the system of interest from her point of view. Once an architect decides to serve a stakeholder with an architecture description, she needs to understand the related point of view and capture it in an architecture viewpoint. An architecture viewpoint governs the creation of one architecture view. The architecture viewpoint serves as communication element explaining what will be covered in the related architecture view and in which form will it be presented.

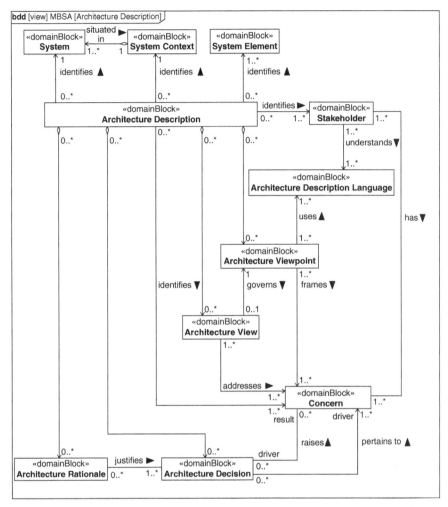

Figure 6.2. Architecture descriptions comprise views and viewpoints, architecture decisions, and related rationales. It identifies the system, the system context, stakeholder(s) and their concern(s), and system element(s) as building blocks of the system.

This imposes that the architecture viewpoint will exist before and hence can exist without the related architecture view.

An architecture viewpoint frames one or more concerns. The architecture viewpoint defines one or more architecture description languages to be used while describing system aspects that relate to the framed concerns. This means, an architecture viewpoint brings together architecture description languages with concerns. Consequently, it is only useful for stakeholders that share at least some of the framed concerns and understand the used architecture description languages.

The architecture viewpoint aggregates a number of visualization kinds of the selected architecture description languages that shall be used to create the related architecture view. The standard ISO/IEC/IEEE 42010:2011 [64] designates such conventions for visualization as "model kind". Visualization kinds are parts of the respective architecture description language and define how certain aspects of architecture shall be visualized. Examples include

- definitions on how to build N-square diagrams
- definitions of traceability tables in natural languages
- definitions on how to build structure diagrams such as block definition diagram and internal block diagram in SysML
- definitions on how to build function block diagrams such as IDEF0.

Obviously, the description of an architecture viewpoint will often reference the used visualization kinds rather than describing them in full.

Architecture viewpoints can become a part of architecture frameworks as further described in Chapter 16.

Figure 6.3 depicts the definition for "Architecture Viewpoint."

The SysML specification declares its viewpoint element [105] as consistent with the standard ISO/IEC/IEEE 42010:2011. But it does not explicitly document any mapping between the two definitions. A SysML viewpoint element maps to the architecture viewpoint as defined above. Version 1.3 of SysML provides the attribute "methods" in the «viewpoint» stereotype to capture visualization kinds. This attribute is intended to designate the methods used to construct the related view. A visualization kind can be rated as part of such methods. Version 1.4 of SysML changes the definition of the «viewpoint» stereotype. The "method" attribute becomes derived and will no longer serve to capture visualization kinds. The newly introduced attribute "presentation" defines the formatting and styling of the governed view. Therefore, in SysML version 1.4 the attribute "presentation" of the stereotype «viewpoint» captures the visualization kinds as described in this section. The SysML diagram type could be mapped to visualization kind.

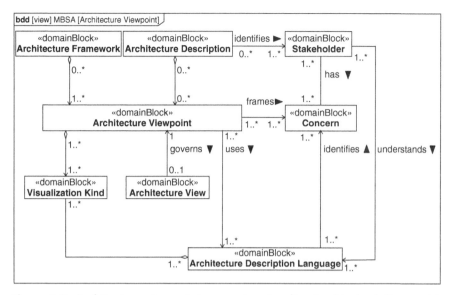

Figure 6.3. Architecture viewpoints comprise visualization kinds, defines architecture languages to be used and frames concerns to govern the creation of architecture views.

6.2.2   Architecture View

Architecture views are an essential part of an architecture description. They visualize the system of interest as it appears from the related architecture viewpoints using the defined visualization kinds. It is built from one or more visualizations to describe elements and relationship among them. Visualizations are the building blocks of architecture views. The standard ISO/IEC/IEEE 42010:2011 [64] designates these building blocks as "architecture models". This is a concept along the line "An illustration is a model of the illustrated item". But the term competes with the model we consider in our model-based system architecture.

A visualization will reveal the system elements, their behaviors, relationships, and interactions. The concept of a visualization as an intermediate entity between visualization kind and architecture view enables reuse. A visualization can be used in one or more architecture views. Visualizations shall identify their governing visualization kind.

An architecture view as a visualization of the system of interest addresses one or more concerns. A concern may be addressed by more than one architecture view. This is the case if a concern cannot be

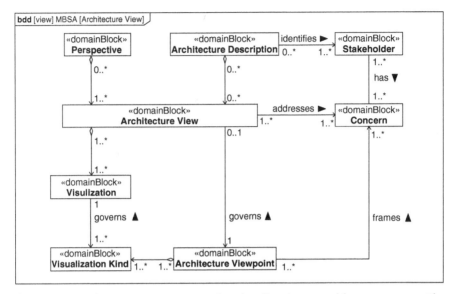

Figure 6.4. Architecture views comprise visualizations to address concerns. The creation is governed through architecture viewpoints. Architecture views may aggregate into perspectives.

addressed completely with one architecture view or because a concern needs to be addressed from different architecture viewpoints. The latter happens if more than one stakeholder share the same concern but need to be served with different architecture description languages.

The stringent definition for the multiplicity of architecture view and architecture viewpoint of 0..1:1 will result in a considerable number of architecture views. They will partly overlap and cover related issues. The "perspective" as introduced by the TRAK architecture framework [116] (see also Section 16.3.7) permits to categorize and group architecture views. An architecture view may belong to any number of perspectives.

Figure 6.4 depicts the definition for "Architecture View".

As with the viewpoint element, the SysML specification declares its view element [105] as consistent with the standard ISO/IEC/IEEE 42010:2011. A SysML view element maps to the architecture view as defined above. SysML diagrams could be mapped to visualizations. Consequently, an architecture view can be expressed with one or more SysML diagrams. And a SysML diagram can represent a part or a whole architecture view.

6.2.3   Architecture Decision and Architecture Rationale

The process of architecting is a sequence of decisions. Documenting these architecture decisions with the justifying rationale is very valuable later in the life cycle of the system of interest. It documents why the system of interest came to its actual design. Stakeholder may easier accept and understand the system of interest when they get reasoning for a certain design. A comprehensive reasoning on architectural design will support impact analysis and minimize risks when deciding on future change requests. Moreover, proper documentation of decisions with appropriate rationale is a good engineering practice.

An architecture decision should be documented at least in cases of dilemmas caused by conflicting requirements or trade studies with no clear favorite. An architecture decision should comprise of what the decision was, who took the decision in which role and at what date. A trace to one or more architecture rationales should justify the decision.

Architects should be aware that architecture decisions do not only pertain to concerns, but may also raise new concerns. An architecture decision may cause new impacts to processes operated by already identified or other stakeholders. And of course, one architecture decision may depend on another architecture decision.

Architecture decisions shall also trace to affected architecture description (AD) elements. The standard ISO/IEC/IEEE 42010:2011 [64] describes the AD element as the most primitive construct of an architecture description. This includes the system elements, their behaviors, relationships and interactions, but also the system, system context, stakeholders, concerns, architecture viewpoints, architecture views, visualization kinds, visualizations, architecture decisions, and architecture rationales. In summary, each entity described in an architecture description is rated as architecture description element.

An architecture rationale should provide the justification for related architecture decision. This is often by applying principles. Principles, such as design or architecture principles (including organization or interaction principles etc.) or evolution principles can be seen as kinds of architecture rationales. Other kinds of architecture rationales include trade studies, simulation results, risk analysis, or other methods providing evidence as basis for a certain decision. A comprehensive rationale becomes most important in case of liability issues. Rationales often capture data that documents core knowledge of a company.

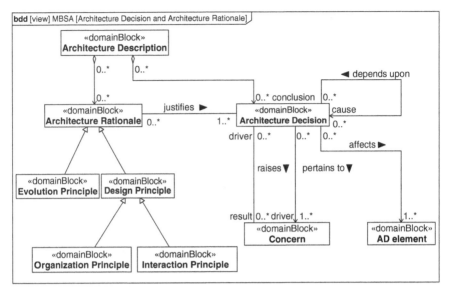

Figure 6.5.   Architecture decision and architecture rationale with their associations.

Figure 6.5 depicts the associations of "Architecture Decision" and "Architecture Rationale."

SysML [105] defines only an element for rationales. There is no dedicated element for architecture decisions. Since SysML is extensible, an architect may define a stereotype «architectureDecision». This would permit to explicitly document and consequently trace architecture decision as described above. Figure 6.6 depicts the definition of such new stereotype and the application of the new element.

6.3   HOW TO GET AN ARCHITECTURE DESCRIPTION?

Pushing the print button and your modeling tool creates all the required documentation. This is the vision many architect may have. Modeling tools, with their heritage in software engineering do not always provide optimized capabilities to create descriptions of system architecture. Model-based approaches are very often limited to certain domains or disciplines. For many types of data exchange documents are still required or even prescribed. We can expect that some of them will remain for a long time from now. A further challenge is, that certain architecture views are used in documents that are not primary

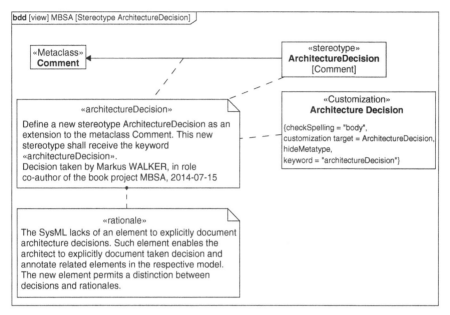

Figure 6.6. Definition of stereotype ArchitectureDecision as SysML extension with its application and rationale.

architecture description. For instance, the standard on requirements engineering ISO/IEC/IEEE 29148:2011 [63] describes five kinds of documents. Four of them, the stakeholder and system requirements specifications as well as the concept of operation and the system operational concept, comprise a section with a view to the architecture of the system of interest. This example shows, that architecture description, or at least part of it needs to be embedded into documents of other disciplines. Therefore, a comprehensive architecture model needs to support documentation in a rather extensive way. That is, it has to provide fragments of the whole architecture documentation in a form supporting the respective disciplines[2] in their document creation.

An interesting approach in that regard had been demonstrated by the Telescope Modeling Challenges Team of the MBSE initiative by INCOSE and OMG [94]. They developed a plug-in to host documentation within a package of the architecture model. The structure in that package defines the structure of the document intended to create. It includes references to elements in the architecture model and provides

---

[2] We refer here to disciplines acting mainly on the same architecture level related to requirements engineering and verification.

containers for additional text. They called it Model-Based Document Generation. Though only rated as proof of concept, it shows a possible way in improving the documentation issue.

6.3.1   Forms and Templates

Several forms and templates for architecture description exist. Very often they originate from software engineering and comprise sections that are not suitable for systems engineering. The working group on moderate complex systems associated to both, the German and Swiss chapter of INCOSE, presented such template at the "Tag des Systems Engineering", the German annual systems engineering conference 2014 [3]. Figure 6.7 depicts the headline structure and Table 6.1 provides explanations on the intended content for this template.

The target was to define a template with less than five pages, sufficient for moderate complex systems. The working group considered a number of sources and fostered for alignment with ISO/IEC/IEEE 42010:2011 [64]. The latter was not considered as regulatory requirement but as descriptive standard relying on experiences. The intention of this template is to get a single document useful for a limited number of stakeholders.

Figure 6.7. Headline structure of the Architecture Template 1.0 by the working group on moderately complex systems of Gesellschaft für Systems Engineering, e.V. (German chapter of INCOSE). Reproduced with kind permission from the German and Swiss chapter of INCOSE.

TABLE 6.1 **Explanation of Content of Architecture Description Template for Moderate Complex Systems**

Heading	Intended Content
1. Introduction	Short explanation on what the document is about.
1.1 Project	Description of the project in which the architecture description was created. This describes the context in which this work product had been created.
1.2 System of Interest	Introduction into the system of interest.
1.2.1 System Context	The system context comprises each entity interacting with the system in a not negligible way. This section shall identify these entities. Consequently, this section defines the system border. ISO/IEC/IEEE 42010:2011 designates the system context as environment.
1.2.2 Purpose of the System	The purpose defines why the system should be built. It may explain why it is beneficial for its users and customers are willing to pay for it. ISO/IEC/IEEE 42010:2011 defines the purpose of a system as specialization of concern.
1.3 Base Architecture	Constraints imposed by stakeholders that limit the solution space. Such constraints may relate to the business case of the company that develops the system of interest. To impose the base architecture is an architecture decision. The related rationale is frequently linked to the business model or habits within companies.
2. Stakeholder, Concern, Viewpoint, View	A list of the considered stakeholders, their concerns and related viewpoints and views. It references the section of the document with the respective view. This is a kind of map of the document to direct each considered stakeholder to the sections of interest. This addresses sequential nature of a document that permits to optimize the sequence for only one viewpoint. One of the simplest representation of this data could be a table. With more than only a hand full of stakeholders, other representations than tables should be considered.

(*continued*)

TABLE 6.1 (*Continued*)

Heading	Intended Content
3. Architecture Description	Description of structure and behavior of the system of interest in views as mentioned above. Each view is captured in a dedicated section. Obviously, this is the core of the document. The structure within this chapter may serve to group related views.
4. Architecture Decisions	Each architecture decision assigned with a unique identifier, for instance in form of a numbered list. Each decision should be linked to the name of the persons that did take the decision and their roles at the date of the decision as well as to a comprehensive rationale for that decision.
5. Explanations	Description of vocabulary specifically used in this document. This refers to the used architecture description languages and could also comprise references to the related glossaries or standards.
5.1 Glossary	For natural languages, a list of terms with related explanations that are important to understand the content of the document.
5.2 Used Symbols	For graphical languages, a list of used symbols with the related explanations.
6. References	A list of referenced artifacts.

Another form was presented at EMEASEC 2014 by Beasley et al. [12]. They proposed a structure in analogy to the structure of well-established requirements documents. This approach supports readers that previously read the requirements document and eases to follow the transformation from requirements to architecture.

Chapter 7

Architecture Patterns and Principles

Did you know that you use patterns every day? It is a very powerful mechanism and much more powerful when they are explicit and known. A pattern describes a proven solution for a problem. Every experienced engineer has a huge set of patterns. Think about your daily work. If you have to solve a problem that you have solved successfully before, you will remind your solution and apply it again. If you describe your solution in a general and reusable way, you have made your experience explicit and persistent. Voilà! You have found a pattern.

Architecting is not a task that can be done simply with checklists and predefined processes. It is strongly based on the experience and talent of the system architects. Patterns are good tools to make the implicit knowledge of the system architect explicit. Nevertheless, checklists and processes still support the architect and are also in one drawer of her toolbox.

Patterns and principles are documented best practices and proven experiences of system architects. It is recommendable to consider them and you need very good reasons to work contrary.

You can't invent a pattern. You can find and describe a pattern. One of the first who did that and triggers the pattern community with his work "A pattern language" was C. Alexander [5]. His book is not about

Model-Based System Architecture, First Edition.
Tim Weilkiens, Jesko G. Lamm, Stephan Roth, and Markus Walker.
© 2016 John Wiley & Sons, Inc. Published 2016 by John Wiley & Sons, Inc.

software or systems engineering patterns, but about architecture patterns for buildings. Of course, manifesting best practices by writing them down works for every discipline.

This chapter lists architecture patterns and principles that we find useful and which we reference from other chapters in this book. The list is not complete and you should identify your personal set for your architecture toolbox. For more information about patterns and systems engineering we recommend the paper "Applying the Concept of Patterns to Systems Architecture" from Cloutier and Verma [21].

7.1 THE SYSMOD ZIGZAG PATTERN

The pattern is described as part of the SYSMOD methodology in Refs [147, 145]. It describes the relationship between requirements and architectures on different abstraction levels.

Requirements specify what the system should do and the system architecture and design how the system satisfies the requirements. This rule seems to be easy. However, on the second view it will turn to be a challenging question. What is part of the architecture and what is part of the requirements?

Let's assume you have absolutely solution-free requirements (although most of those requirements are not viable in practice). Now you derive a system architecture that satisfies the requirements and you get the typical what/how-pair. For example, consider the requirements of our guided virtual live tour through a real museum. You derive a system architecture with mobile museum robots that satisfies the given requirements (figure 7.1). That solution leads to new requirements that contain aspects of the solution, for example, requirements about collision detection and avoidance of the museum robots with things or people in the museum. You won't need those requirements when you

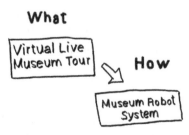

Figure 7.1. What and how.

have derived a system architecture for instance without moving robots, but with a static camera system.

The requirements depend on a given architecture and are not free of any solution. They are on another abstraction level and solution-free from the viewpoint of that level, but they contain solution aspects of the previous level.

Again you derive a solution from the requirements, for example, a camera-based system for collision detection. That solution leads to new requirements, and so on (Figure 7.2). All in all the logical steps represent a zigzag pattern. "Zig"—from requirements to a system architecture—and "Zag" back from the architecture to the what-side of the derived requirements. Then again a "Zig" from the derived requirements to the architecture, and so on.

The zigzag pattern also shows the relationship between the different requirement specifications and architecture kinds (Figure 7.3). The "Stakeholder Requirements Specification" is the top-level specification in a project. It specifies the requirements of stakeholders. Next, a system architecture that satisfies the requirements is derived. It is a logical architecture that specifies the technical concepts but omits the technical details. That architecture is the base for the "System

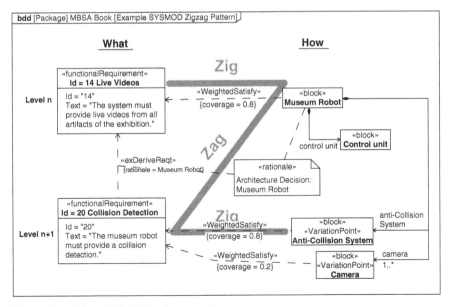

Figure 7.2. The SYSMOD zigzag pattern.

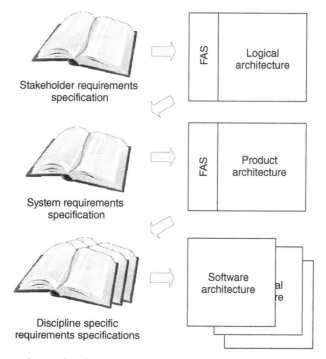

Figure 7.3. Relationship between requirements and architecture kinds.

Requirements Specification". While the stakeholder requirements specification" defines what the system should do from the stakeholders' perspective, the system requirements specification defines how the system should satisfy the stakeholder requirements from the engineers' perspective.

Next the product architecture is derived from the system requirements specification. The product architecture satisfies the system requirements and includes all details necessary on the system level to build the system.

On the next level, you can derive the discipline-specific requirements for the software, mechanical, electrical, and so on assemblies. The functional architecture exists on each level as another architecture kind.

Probably that reminds you of the V-Model.[1] The zigzag pattern covers the left branch of the V-Model as shown in Figure 7.4.

---

[1] Readers not familiar with the V-Model can find its description and a short summary of its history in Appendix B.

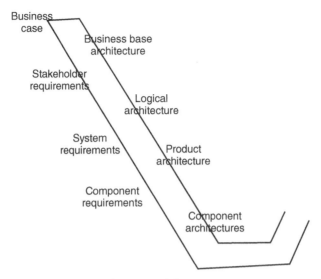

Figure 7.4. Zigzag pattern in the V-Model.

A common problem in practice is a strong focus on the left side of Figure 7.3. Stakeholders, principals, and others communicate and discuss mainly the requirements and not the architectures behind them. The requirements and their architecture one level above — the so-called base architecture — should be treated as one artifact in the engineering process. See also Section 7.2 for more information about base architectures. There are also projects with a strong focus on the right side, that is, projects where the requirements are more in the background and the development is driven by the technology. In our opinion, the best option is as always to steer a middle course.

The zigzag pattern is independent of SysML or any other requirements and architecture modeling languages or methodologies. It is a generic pattern that describes the relationship between requirements and architectural elements across abstraction levels. A similar approach is the function-mean tree developed by Andreasen [9] and the Axiomatic Design [132]. Another one is the Systems Engineering Sandwich described by Dick and Chard [28]. All of them describe the co-evolution between functions and solutions.

If you want to model the zigzag pattern explicitly with SysML you need the following relationships and model structures: According to the model structure template given in Section 7.9, the system requirements are located in a top-level package called "<system>_Requirements."

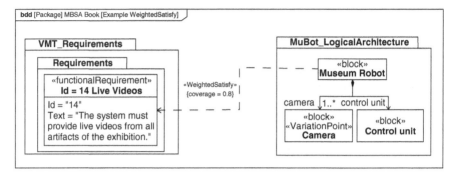

Figure 7.5. One zigzag pattern level in the model.

Typically, with subpackages to further organize the requirements. The architecture elements that satisfies the requirements are located in a package "<system>_LogicalArchitecture" or "<system>_ProductArchitecture." Again there are further subpackages to organize the architecture. Elements of the architecture have satisfy relationships to the appropriate requirements. This structure models one level of the zigzag pattern (Figure 7.5).

On the next level the requirements have derive relationships to one or more requirements from the upper level. They are based on architecture decisions made in the architecture. The SysML derive relationship itself has no property to store the rationale why and how the requirement was derived. We present two options to relate the rationale with the architecture elements to the derive relationship:

1. SysML provides the rationale element. A special comment to document the rationale why something was modeled like it is. Attach a rationale to the derive relationship and model anchors (dashed lines) from the rationale to the architecture elements (Figure 7.6).
2. Introduce a new stereotype for an extended derive relationship that specializes the SysML stereotype *«deriveReqt»* and adds a property to store the information about the architectural elements that lead to the derived requirements (Figures 7.7 and 7.8).

The best option depends on your needs. The first one has the most explicit visualization of the relationships. The second option could easily be accessed by tools to analyze the relationship.

Figure 7.9 shows a composition relationship from the museum robot block to the anticollision system. That relationship crosses the border

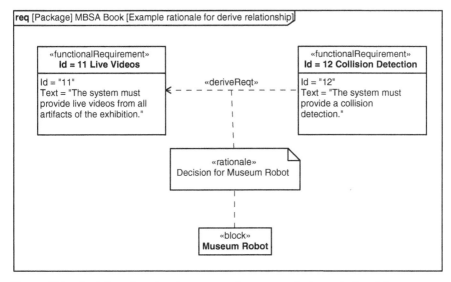

Figure 7.6. SysML rationale element to document derive relationship.

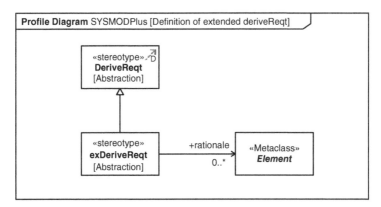

Figure 7.7. Definition of extended derive relationship.

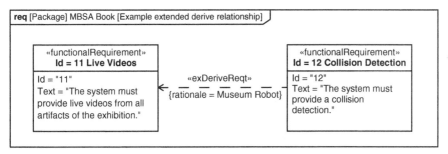

Figure 7.8. Example of the extended derive relationship.

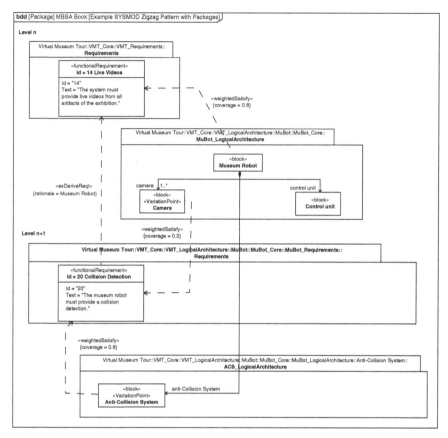

Figure 7.9. Model structure and relationships of the zigzag pattern.

of the zigzag levels in the wrong direction. The lower level must depend on the upper level. Here the upper level depends on the lower level. In the strict sense that is not allowed. However, as mentioned earlier, it is not necessary to strictly model the zigzag pattern. It costs effort to strictly separate the levels and the effort should only be spent if you gain the benefit. To strictly separate the elements of the zigzag levels you can introduce a specialized version of the museum robot as shown in Figure 7.10.

The zigzag pattern has value in communication and modeling. Even if you don't model the pattern structure explicitly, it demonstrates the relationship between requirements and the system architecture and the abstraction levels. It gives you a vocabulary to structure discussions about requirements and to conduct them in a constructive spirit.

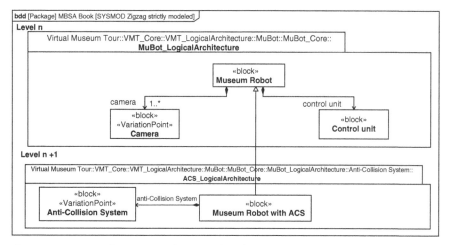

Figure 7.10. Strictly separated zigzag levels.

Equipped with the correct vocabulary, you can for instance express the decision which levels of requirements or architecture to cover with a SysML model and which ones to omit during the modeling.

If you model the pattern explicitly, it provides traceability paths from the system architecture to the requirements over many abstraction levels. It clearly shows the separation of requirements engineering and system architecting and at the same time shows the need for a close collaboration between the requirements engineers and the system architects, which will be discussed in detail in Section 10.2.

7.2 THE BASE ARCHITECTURE

Your system requirements typically do not start at the very top of the zigzag pattern levels (Section 7.1), that is, they already include some technical decisions. An aviation company has requirements about planes and an automotive company about cars, and not about transportation systems in general.

Some of the architectural decisions are obvious, some are implicit. They are one of the causes why requirements are a sore spot of many projects. The knowledge and assumptions about implicit architecture decisions is different between the project members.

The zigzag pattern shows that the architecture decisions that are the input of the engineering project are described in the architecture

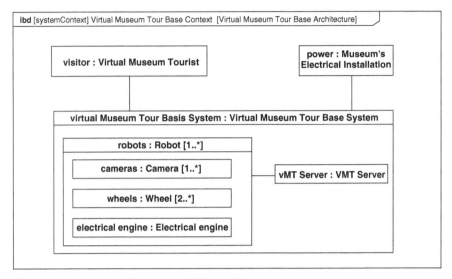

Figure 7.11. Base architecture of the VMT.

one level above the top-level requirements. From the viewpoint of the requirements that architecture is called the *base architecture*. The following figure shows an extract of the base architecture of the VMT (Figure 7.11).

The VMT base architecture clearly specifies that the system receives power from the museum's electrical installation, it will have robots with wheels, an electrical engine, and cameras that are controlled by a central control server. The top-level requirements of the VMT depend on the base architecture and directly use those elements. They are only solution-free according to their level in the zigzag pattern. For instance, we have top-level requirements "Robot Distance" and "Robot Mass" that are based on the decision to use robots. We won't have those requirements if we had decided for a static camera system. Figure 7.12 shows the dependency of the requirements to the base architecture.

The base architecture constrains the solution space and makes implicit technical decisions explicit. Therefore, it is also a good place to spot potential for disruptive innovations. You can put a question mark over the decisions in the base architecture.

In practice, a base architecture description is a set of block diagrams and additional textual descriptions. At minimum it is a context diagram, a product tree, and an architecture block diagram. Figure 7.11 depicts a context diagram and an architecture block diagram. Figure 7.13 depicts

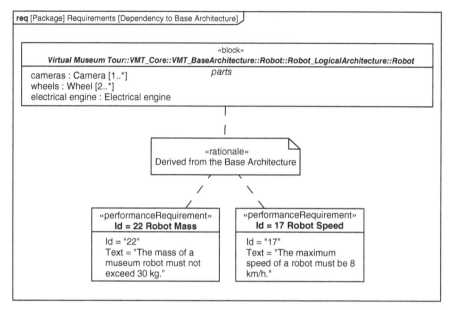

Figure 7.12. Relationship of top level requirements to base architecture.

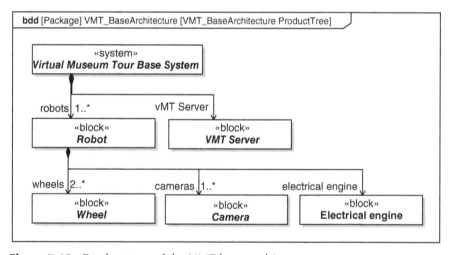

Figure 7.13. Product tree of the VMT base architecture.

the product tree. For each actor and part of the architecture, a textual brief description is provided in a table (Table 7.1).

You could also formalize the base architecture with constraint requirements. The base architecture provides technical constraints

TABLE 7.1   Brief Description of Base Architecture Elements of the VMT

Property	Brief description
robots:Robot[1..*]	The robots are physical systems that are controlled by the "VMT-Server" or the "Virtual Museum Tourist." The robots move on wheels and use cameras.
cameras:Camera[1..*]	At least one camera to record the environment of a robot for the users and to navigate.
wheels:Wheel[2..*]	At least two wheels to move the robot through the museum.
vmtServer:VMTServer	One server to control all robots of a single VMT.
visitor:Virtual Museum Tourist	The visitor of the VMT is a human user.

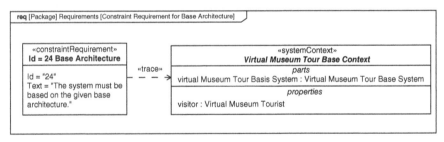

Figure 7.14. VMT constraint requirement for base architecture.

for the system under development. Depending on your requirements engineering process you could consider to transform the constraints of the base architecture to constraint requirements in your requirements model. Otherwise, you must ensure that the base architecture is always part of your requirements model and documentation.

A pragmatic approach is a single constraint requirement with a trace relationship to the root element of the base architecture (Figure 7.14).

Many projects do not explicitly describe the base architecture. It is implicitly described in the requirements and leads to misunderstandings, that is, errors, effort, and costs. It is simple to avoid that obstacle in your project by documenting the base architecture.

7.3 COHESION AND COUPLING

Cohesion and Coupling is a common applied principle for systems. Strictly speaking, these are two principles that act contradictory (Figure 7.15).

Cohesion Principle

A part of a system should be cohesive as possible, that is, the means of the part are closely related.

Simply spoken, the cohesion principle says that things which do similar tasks should be at the same place. Short ways make them more effective and changes have local effects.

Coupling Principle

A part of a system should be loosely coupled as possible, that is, it has a minimum number of weak dependencies to other parts of the system.

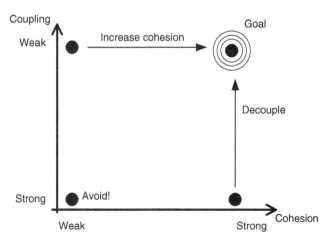

Figure 7.15. Cohesion and coupling (Copyright oose Innovative Informatik eG).

A dependency between parts could be explicit or implicit. Parts are explicitly coupled for instance if they have a mechanical or an IT data exchange connection. The coupling is stronger the more likely it is that a part must be updated when the connected one is changed.

Parts are implicitly coupled when they have a logical dependency without having an explicit connection. For instance if a behavior implemented in a software depends on a physical parameter of a part. These hidden links are critical since they are hard to identify and could lead to undesired behavior and malfunction of the system when one part is being updated without the other one.

The notion of the coupling principle is to make the system more modular. Parts could be updated or replaced without having a (strong) impact on other parts.

The principles cohesion and coupling act contradictory. If you increase cohesion, you also increase the coupling. A strong cohesion leads to more parts with each having only a small set of the overall provided functions. Less functions fit more to a strong cohesion criterion of a single part. In consequence, you have more parts to cover all functions. More parts of the system require more relationships, that is, more coupling.

If you decrease coupling, you also decrease cohesion. The extreme case is a system that consists of a single assembly. The coupling on that level of detail is zero. But the cohesion is extremely weak, because that assembly provides all functions of the system even if they are not closely related and have nothing in common.

Note that the cohesion and coupling principle appears on different levels of the architecture. The single assembly in our example above contains parts and those parts could have a strong coupling and cohesion on their level of detail. And each part of the parts could again include parts and so on. The cohesion and coupling is different on each level.

The software engineering discipline is the origin of the cohesion and coupling principle. Larry Constantine published the metrics in his article "Segmentation and Design Strategies for Modular Programming" in 1968 [25]. The principles are also known in nonsoftware disciplines. Karl Ulrich covers the coupling of mechanical interfaces and related dependencies between the parts in the paper "The role of product architecture in the manufacturing firm" [139].

7.4   SEPARATION OF DEFINITION, USAGE AND RUN-TIME

The definition of elements and the definition of their usage in a specific context and the definition of their structure and relationships at run time are separate aspects and should be handled separately.

The definition of an element is the blueprint, that is, it defines the structure and behavior. A simplified example is a screw. You can define the head, shaft, material, length, diameter, and so on of the screw. For common elements like screws, the definition could be published in a parts catalog. It is the blueprint for all screws of that type and not a representation of a concrete single screw from the real world. In SysML, we specify the screw definition with blocks in a block definition diagram (Figure 7.16).

The usage defines the application of a screw in a specific context. For example that our screw is used to mount the image acquisition unit on the housing of the museum robot. The definition of the screw constrains the usage level. You can only use the elements according to their definition. The other direction the usage level has no impact on the definition. In SysML, we define the usage within a context with the internal block diagram (Figure 7.17).

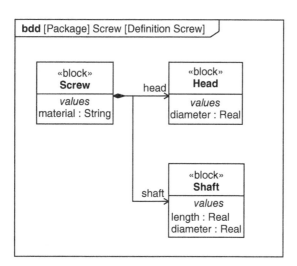

Figure 7.16. Definition of a screw with SysML.

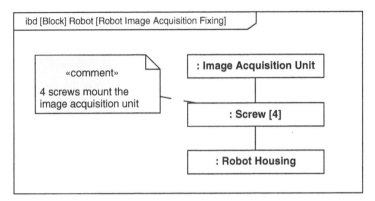

Figure 7.17. Usage of a screw with SysML.

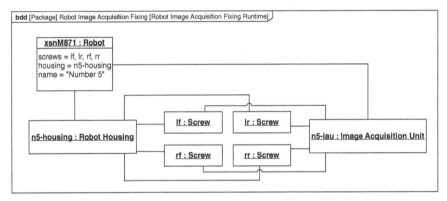

Figure 7.18. Runtime setting of the screw with SysML.

The run-time level specifies the links and properties of elements at a specific point of time during the run time of the system (Figure 7.18).

Be aware of the different levels. It is easy to mix them up. For example, the usage level could also be part of the definition of an element. Figure 7.19 shows the internal block diagram of the screw. It specifies the usage of the head and the shaft in the context of the screw.

SysML follows the principle of the separation of definition, usage and runtime aspects. The block definition diagram defines the blocks and the internal block diagram specifies the usage in a defined context. Activities define behavior, call behavior actions the usage of the activity, and so on. It is important to know the three levels to handle them separately and to know the relationship among the levels.

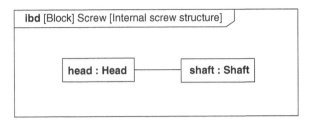

Figure 7.19. Internal structure of the screw.

7.5 SEPARATE STABLE FROM UNSTABLE PARTS

Any model, documentation, or real system has stable and unstable parts. The properties stable and unstable refer to changes. A stable part is seldom changed and an unstable part is often changed.

It is the best practice to separate stable parts from unstable parts. The stable parts should not depend on an unstable part. Otherwise, they adopt the instability.

In the model of a physical product architecture you have stable technical concepts and unstable concrete physical parts. If you follow the principle to separate stable from unstable parts you model an explicit logical architecture (technical concepts, stable) and a product architecture (concrete physical blocks, unstable).

Another example is the separation of concerns using layers. See Section 9.5 about layered architectures.

Note that it is a principle and not a strict rule. Only follow the principle if you gain a benefit. In any case it costs effort.

7.6 THE IDEAL SYSTEM

The ideal system is a tool to narrow the solution space to find a system that really satisfies the user's demands. The ideal system fulfills the requirements of the user without even existing. That sounds ridiculous, but this line of thought is really helpful. Reflections on this obviously unattainable goal restrict the search space for a solution and turn the focus of the system development in a successful direction.

For this purpose, a concrete example taken from Refs [147, 145]: Consider the locking system of a car. Let's start at the time when it was necessary to put a key into the lock and to turn it to unlock a door of the car. Is it that what you want as a user? With all the lovely side effects that the lock is frozen in winter or the paint of the car is being scratched from the key. As a user you don't want all these. You want to protect your car against theft and misuse.

Today's cars have a central locking system, which can be opened with a remote control on the key. One click locks or unlocks the doors. The locking system is less existing for the user and provides the same functionality as the key-based system. It is still not an ideal system and has some inconveniences. For example, you notice after fastening the seat belt that the key is still in your pocket and you need it to start the car.

Latest technology already is very close to the ideal system. Once you use the door handle of the car the car recognizes whether an RFID key is nearby and opens the car when it is detected. With a fingerprint scan on the shift lever the motor can be started. For the user the system is barely existent, but the desired functionality is available.

Consider the systems that surround you and think about how these systems evolved over the years. Mostly they strive toward the ideal system. The principle is valid for most systems. The user requests functionality and not the system itself.

The guiding principle is: Focus on solutions that are as inexistent for the user as possible.

The ideal system is a principle from the TRIZ methodology [7]. TRIZ is an acronym that stands for теория решения изобретательских задач (триз), which is Russian and means "Theory of Inventive Problem Solving" (TIPS). This theory is based on a systematic methodology for innovative processes. The father of TRIZ—Genrich Soulovich Altshuller (*15.10.1926 – †24.09.1998)—was convinced that inventions are not coincidences. He has analyzed thousands of patents and derived TRIZ from his insights. The ideal system is one pattern he has observed.

7.7 VIEW AND MODEL

An important principle in modeling is the separation of view and model. The model is the source of the information. The modeling language defines the data structure and the semantic of the model elements. The data structure is also called abstract syntax. The left side in Figure 7.20 shows the abstract syntax of the SysML use case and includes relationship model elements. The abstract syntax is not a notation but a definition of the data structure. The storage format could be XMI, an XML language for SysML and UML models [109].

The view is a textual or graphical representation of the model. Typically, a modeling language provides a notation for the model

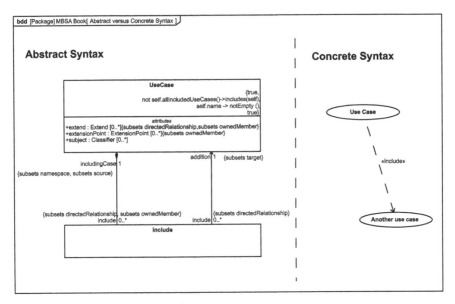

Figure 7.20. Abstract syntax of a use case.

elements—the so-called concrete syntax. The right side in Figure 7.20 depicts the concrete syntax of the SysML use case and include relationship. However, not every modeling language provides a view. For instance, the "Business Motivation Model" (BMM) [101]—a standard of the OMG for modeling vision, mission, strategies, goals, and more for businesses—does not provide a notation and only the abstract syntax and semantics.

The view should not contain any additional semantic that is not part of the model. That means you can remove all views without losing relevant information. See Section A.1 about the separation of view and model in SysML.

If your primary focus of modeling is the communication between the development team members, the view is more important than the model. You can even create views without a real model like flipchart sketches or drawings in slideshow applications. That is not decent modeling but more painting. If the specification, analysis, or simulation is your primary focus then the model is more important than the view. You could even discard the view and work only with the model.

Naturally, people look on the view artifacts and not the model. In modeling environment like SysML, the view is also the editor to create or change the model elements. The "real" model could easily get out of

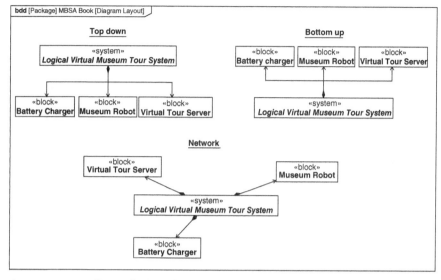

Figure 7.21. Different layout directions.

the scope. Be aware of it and handle the view and the model according to your requirements.

7.8   DIAGRAM LAYOUT

In system modeling, the graphical representation of the model data is an important part. It definitely makes a difference how you present your data. The human brain looks for graphical patterns and not for all the little details. That makes life much easier. For example, you can change the letters in a text while keeping the pattern structure and you could still read the text:

"Mdoel Baesd Setmsys Ennneigierg mekas my procjets mroe ecffetive."

Similar things happen when you look at model diagrams. You'll see graphical patterns. Unfortunately, such patterns are not well known and described. Here's a huge potential to be more effective. For example, it makes a difference if you layout a product tree in a block definition diagram top-down, bottom-up, or in a network layout style (Figure 7.21). We prefer the top-down layout. It emphasizes the breakdown of the product (Product Breaddown Structure (PBS)).

Some notes about diagram layouts:

- In the authors' culture, the reading direction is from top to down, from left to right. It is different in other cultures. Layout your diagram conform to the usual reading direction of your stakeholders.
- A diagram should not contain more than 7 ± 2 major elements [96].
- A diagram should have a printable layout (A4 or letter format).
- Rarely use colors in your diagrams. Some people are color-blinded, and monitors, printers, and beamers present colors differently. Instead use gray color to emphasize specific aspects, if you need different colors in your diagram.
- One diagram has one purpose. A new purpose requires a new diagram.
- Look through the eyes of the model reader when you create a diagram and not through the eyes of a model builder.
- Clearly separate diagram creation from model building. See also Section 7.7.

7.9 SYSTEM MODEL STRUCTURE

We will look at system model structures from the viewpoint of a SysML model. Although the basic concepts are valid for all system model kinds.

At first glance, it seems to be simple to define the package structure of a SysML model. However, you will often get problems with an implicitly built structure. A model has many – orthogonal – aspects and abstraction levels that could be mapped into the package structure, for instance domain, modeling, or organizational aspects. You can easily mix up the orthogonal aspects. You probably know the problem by organizing your hard disk with a sophisticated directory structure.

The "MBSE Challenge Team SE^2 for Telescope Modeling" describes the best practice for the package structure in the "MBSE Cookbook" [54]. Figure 7.22 shows the top level of the system model package structure.

The root package represents the complete system model. If you model variants (see Chapter 15) the next level has three packages: one for the configurations, one for the core, and one for the variations. If you do not model variants of your system, you skip the package structure of this level and directly proceed with the structure that is

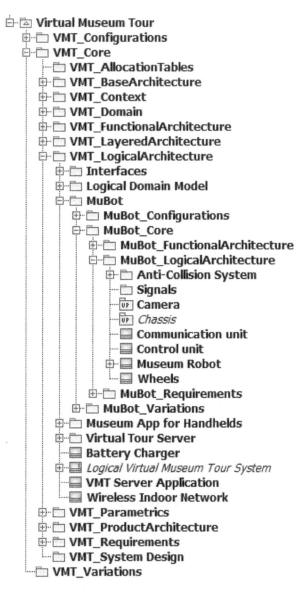

- ⊟ 🖿 **Virtual Museum Tour**
  - ⊞ 🗀 **VMT_Configurations**
  - ⊟ 🗀 **VMT_Core**
    - 🗀 **VMT_AllocationTables**
    - ⊞ 🗀 **VMT_BaseArchitecture**
    - ⊞ 🗀 **VMT_Context**
    - ⊞ 🗀 **VMT_Domain**
    - ⊞ 🗀 **VMT_FunctionalArchitecture**
    - ⊞ 🗀 **VMT_LayeredArchitecture**
    - ⊟ 🗀 **VMT_LogicalArchitecture**
      - ⊞ 🗀 **Interfaces**
      - ⊞ 🗀 **Logical Domain Model**
      - ⊟ 🗀 **MuBot**
        - ⊞ 🗀 **MuBot_Configurations**
        - ⊟ 🗀 **MuBot_Core**
          - ⊞ 🗀 **MuBot_FunctionalArchitecture**
          - ⊟ 🗀 **MuBot_LogicalArchitecture**
            - ⊞ 🗀 **Anti-Collision System**
            - 🗀 **Signals**
            - 🔲 **Camera**
            - 🔲 *Chassis*
            - ▦ **Communication unit**
            - ▦ **Control unit**
            - ⊞ ▦ **Museum Robot**
            - ▦ **Wheels**
        - ⊞ 🗀 **MuBot_Requirements**
      - ⊞ 🗀 **MuBot_Variations**
      - ⊞ 🗀 **Museum App for Handhelds**
      - ⊞ 🗀 **Virtual Tour Server**
      - ▦ **Battery Charger**
      - ⊞ ▦ *Logical Virtual Museum Tour System*
      - ▦ **VMT Server Application**
      - ▦ **Wireless Indoor Network**
    - ⊞ 🗀 **VMT_Parametrics**
    - ⊞ 🗀 **VMT_ProductArchitecture**
    - ⊞ 🗀 **VMT_Requirements**
    - 🗀 **VMT_System Design**
  - 🗀 **VMT_Variations**

Figure 7.22. Top level system model structure.

used in the core package. See Chapter 15 about how to model variants and how to manage the configuration, core, and variation packages.

On the next level in the core package, we separate the different modeling aspects like system context, requirements, logical architecture, and so on. The list in the figure is not complete. You will have your own appropriate aspects for your projects. The prefix of each package refers to the enclosing namespace. Typically, you have many packages of the same name in the model, for instance "Requirements." The prefix shows the context of the specific package.

The logical architecture package contains the structural elements of the systems, that is, the physical blocks. Each block that has a detailed description has it's own package on the next level, for instance, the subsystem "MuBot" (Museum Robot). You treat the package like the system root package and create the same package structure inside. The package "MuBot_Requirements" in the figure below contains all requirements that directly relate to the subsystem "MuBot." Again there is a logical architecture package that contains further packages with the same structure. If you strictly separate all MuBot-specific elements, the MuBot package is a complete (sub-)model of its own for the museum robot including requirements and the architectures.

The package structure is straightforward. It works for models of any size and gives the model builder and users a good orientation. The structure could be used for models of any size. The model of a telescope system from the MBSE Challenge Team SE ^ 2 is a good example of the application of this concept [54].

7.10 HEURISTICS

7.10.1 Heuristics as a Tool for the System Architect

Heuristics are an important tool for the system architect. This has been pointed out very thoroughly by Maier and Rechtin [92], who explain that heuristics are condensed experiences, phrased in an easy way that allows communicating them to others. In that sense, we would like to contribute our own set of experiences phrased into heuristics. But we will not even try to cover the aspect of heuristics in the same proficiency as Maier and Rechtin, because this attempt would most certainly fail.

We highly encourage system architects to make use of the powerful tool that heuristics are, in contrast to pure equations and other deductive findings that may solve well-defined problem statements but could easily fail in addressing the multidomain challenges encountered in systems architecting. Collect your own experiences and condense them into heuristics. Ensure that the heuristics are available when needed. To all those who like to know more, we highly recommend the book "The Art of Systems Architecting" by Maier and Rechtin [92]. For those who are familiar with heuristics in general, we now provide our very own Heuristics:

- *Make it lovely not weird*: Architecture descriptions should be nice to work with and should look good. Sometimes, it is just the considerations about diagram layout from further above that will make this happen. Sometimes, it is also the use of illustrative pictures or the review by a colleague with a good eye for aesthetics that will make this happen. Our experiences with both succeeding to convince people with nice architecture descriptions and failing with too user-unfriendly ones tell us that the effort spent in making communication materials look nice will often pay back.

- *Abstraction cuts the Gordian Knot*: Master complexity with abstraction. If the system gets larger or more complex, then the level of abstraction of the top-level architectural views has to be increased. There is only a limited amount of complexity an architecture core team can deal with. A possible way out is to choose an abstraction level as high as the system's complexity can still be governed on that level.

- *Choose the upper deck:* If in doubt whether to model on a higher abstraction level or a slightly lower one, always choose the higher abstraction level (this heuristic is also known as "Jesko's law").

- *The modeler should stick to ending fast*: Never model so deep that the content of system architecture is redundant with content of engineering documentation of the engineering disciplines. The cobbler should stick to his last, and the architect should stick to the system level.

- *Go where the money is:* Make architectural developments and refactoring of existing system architecture in projects with enough resources and not outside any market-driven activity. The projects whose business case predicts high profit also have the resources to realize architectural improvements. By contrast, an architect sitting alone in the corner of a department may find genius architectural improvements but may lack the lobby for getting them implemented.

7.10.2 Simplify, Simplify, Simplify: Strenght and Pitfall

At the end, let us come back to Maier and Rechtin [92] once more. One very central heuristic they mention is "**Simplify. Simplify. Simplify.**" (which is very close to the commonly encountered "keep it simple").

This heuristic applies in many contexts and should often be the driver of the system architect's action. If there is a choice between a simple and a complex concept, then the simple one is most of the time the better one. If a simple sketch of the problem is sufficient to solve the problem, it is not necessary to make a more detailed or more complex sketch about the problem. For example, if an envisioned new product is known to have 50% more interfaces to external systems than its predecessor, then it is probably not necessary to make the detailed interface specifications to find out that the system architecture work-load for the given product will be higher than one of its predecessor. The system architect in charge of initiating a rough effort estimate should thus maybe start making a list of all interfaces rather than diving into one or two of them.

The "Simplify. Simplify. Simplify." comes with one pitfall however: If a system is complex, then it should be handled by simple means, but not by too simple means. Systems become complex when they have to satisfy complex requirements. The complexity has to be controlled somewhere. The strength of model-based systems architecting is the creation of different views on the same complex system—each of them showing only one aspect and thus hiding the complexity for the stakeholders using this view. The model however corresponds to the complexity of the real system to a degree that is sufficiently close for solving the problems that arise from the complexity.

Models of complex systems will probably become complex as well. While the system architects have to govern the complexity, the stakeholders have to focus on the aspects of the system that are relevant to them via the appropriate views. They have to trust the system architects that they will control the overall system and have to be able to live with their limited view on the system. Critics claiming that the system architects are creating too much complexity are often not following this principle of trust. They like to understand the full model themselves, which is often no longer possible with complex systems.

Often mentioned quotes carry a lot of truth: "Everything should be made as simple as possible, but not simpler." (Albert Einstein) [23].

Chapter **8**

Requirements and Use Case Analysis

The requirements and use case analysis are not part of the system architect's activities. However, the architect has close interfaces to that discipline, for instance, when working with functional architectures (Chapter 14). In this section, we give a brief description of the requirements and use case analysis. We follow the "Systems Modeling Toolbox" (SYSMOD) for SysML [145]. The methodology uses common methods of requirements and use case analysis and is not specific for a modeling tool. The core tasks are as follows:

1. Identify and define requirements
2. Specify the system context
3. Identify use cases
4. Describe use case flows
5. Model the domain knowledge.

Hereinafter, we briefly describe each task with a focus on the interface to the system architecture discipline. The architecture tasks of SYSMOD are not covered in this chapter.

The interface between requirements engineering and system architecture is often underestimated in system development. Typically

Model-Based System Architecture, First Edition.
Tim Weilkiens, Jesko G. Lamm, Stephan Roth, and Markus Walker.
© 2016 John Wiley & Sons, Inc. Published 2016 by John Wiley & Sons, Inc.

in focus is the derivation of the architecture from the requirements. That requires a close communication between the architect and the requirements engineer to resolve unclear and conflicting requirements.

In addition, as shown in the zigzag pattern (Section 7.1), the technical decisions of the architect lead to new requirements. Altogether the requirements engineers and system architects must closely collaborate and should not just communicate via documents and models.

Since requirements and architecture are closely related and artifacts are linked, we recommend to have the linked artifacts in the same model. Elements of the architecture are linked to requirements they satisfy and requirements are linked to elements from that base architecture that have led to those requirements.

Figure 8.1 (a) and (b) give an overview about the main artifacts of the requirements analysis tasks.

8.1 IDENTIFY AND DEFINE REQUIREMENTS

The identification and description of requirements is a wide topic. It is about the elicitation of requirements, the structuring of requirements, documentation of requirements, the wording of requirements, the management of requirements, and so on. We will not cover that here and assume that you have your own processes to handle your requirements in your projects. Other books cover requirements engineering in detail, for instance [122].

In our book, we use SysML as the modeling language to specify the system model and in particular the system architecture. SysML supports the modeling of requirements (Section A.4). There are three noteworthy scenarios for the linking of requirements and the system architecture in a SysML environment:

1. The requirements are modeled with SysML and part of the same physical model as the architecture. The elements are linked using SysML elements like the satisfy relationship. There is a traceability path from the architecture to the requirements.

2. The requirements are modeled outside the physical SysML model of the architecture for instance stored in a separate repository managed by a requirements management tool. You have an adapter to transfer information about the requirements from their original model to the system model in SysML. The requirements in SysML

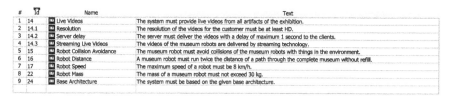

#	Id	Name	Text
1	14	Live Videos	The system must provide live videos from all artifacts of the exhibition.
2	14.1	Resolution	The resolution of the videos for the customer must be at least HD.
3	14.2	Server delay	The server must deliver the videos with a delay of maximum 1 second to the clients.
4	14.3	Streaming Live Videos	The videos of the museum robots are delivered by streaming technology.
5	15	Robot Collision Avoidance	The museum robot must avoid collisions of the museum robots with things in the environment.
6	16	Robot Distance	A museum robot must run twice the distance of a path through the complete museum without refill.
7	17	Robot Speed	The maximum speed of a robot must be 8 km/h.
8	22	Robot Mass	The mass of a museum robot must not exceed 30 kg.
9	24	Base Architecture	The system must be based on the given base architecture.

Systems Requirements

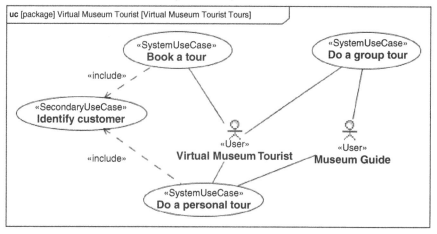

System Use Cases

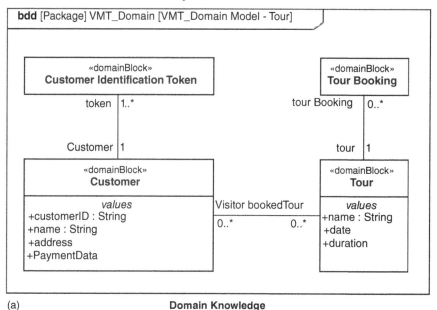

(a)   **Domain Knowledge**

Figure 8.1. (a) and (b) Overview requirements analysis artifacts.

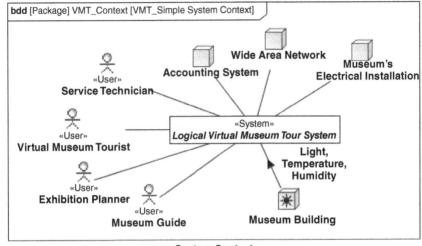

System Context

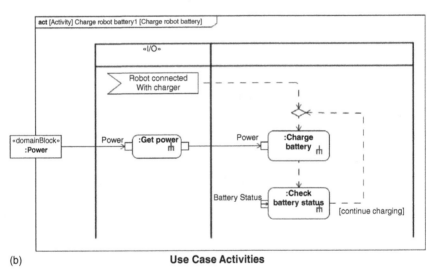

(b)   **Use Case Activities**

Figure 8.1. (*Continued*)

are placeholders for the original requirements in the external physical model. They have SysML relationships to elements of the system architecture. There is a traceability path from the requirements
to the architecture.

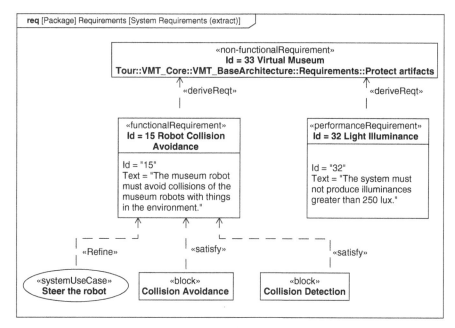

Figure 8.2. Requirements in a SysML model.

3. The requirements are stored outside of the physical SysML model in a model or document. The relationship between architecture elements and requirements is loosely documented in the documentation of the architecture. For instance, a matrix in a textual spreadsheet could be used to manage the relationships. Although there is a traceability path from the architecture to the requirements, the path could not be automatically analyzed since it is not established by model elements.

We prefer scenarios 1 and 2, because the traceability path is completely established by the model.

Figure 8.2 shows a set of requirements in a SysML diagram. The requirements are distinguished between functional and nonfunctional requirements. The nonfunctional requirements are further distinguished between categories like performance or physical requirements.

See Section A.4 about more details about requirements modeling with SysML.

8.2 SPECIFY THE SYSTEM CONTEXT

At the beginning of a system development project, it is necessary to identify the system boundary, external interfaces of the system of interest and interacting systems or humans (e.g., see Ref. [56]). A system context diagram in SysML can be used to describe these elements [145, 147]. Figure 8.3 shows a SysML block definition diagram as a system context diagram, where the black line around the system of interest "Logical Virtual Museum Tour System" depicts the system boundaries, and the solid lines between the system and external entities—the so-called system actors—depicts interactions between the system and its context elements. All further analysis and architecting steps are assumed to describe the system of interest within this identified system boundary.

The actors are classified in different categories: human actors (sticky man) and nonhuman actors (boxes). Figure 8.3 shows also a special kind of a nonhuman actor. The box with the sun represents an environmental effect. Here it is the condition inside the museum building that interacts with our system of interest (temperature, humidity, light, etc.).

Figure 8.3 depicts a simple system context. It is like a graphical list of the system actors. The extended system context shows more details such as system interfaces, actor interfaces, and internal structures of the system of interest and the actors. Typically, it is a task of the system architect to detail the simple system context to a extended system context. Once the information is in the model, the simple and the extended

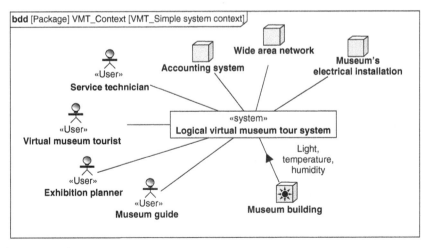

Figure 8.3. System context.

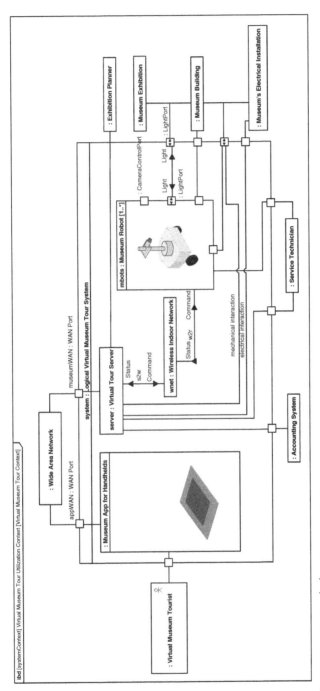

Figure 8.4. Extended system context.

system contexts are different views of the same data. While the simple system context is more suitable for nontechnical stakeholders, the extended system context is for technical-oriented stakeholders like the engineers of the system of interest. Figure 8.4 depicts an extended system context in a SysML internal block diagram (Section A.2.2).

A comprehensive system model has more context specifications than the system context. For instance, the verification context as described in Section 9.8.3.

8.3 IDENTIFY USE CASES

Use cases are an important artifact in systems development. They direct the focus on the user's perspective of the system. The outside-in view supports the development of systems that really satisfies the needs of the users and stakeholders in contrast to the widespread engineering inside-out view of the system. The use case perspective is also well understandable for all stakeholders including people who have no engineering background.

Use cases are a wrapper around the system's functions, defining the pre- and postconditions as well as the trigger of the use case occurring at the system boundary and the result returned to entities of the system context (typically the actor who triggers the use case). Use cases have been extensively discussed in the literature of model-based software and systems engineering [67, 145, 147]. Textual descriptions of system operations [11] or scenarios (e.g., see Ref. [117]) can also be considered as use case descriptions.

Use cases are phrased from the actor's perspective. For example, the use case "Book a tour" is phrased from the perspective of the virtual museum tourist (Figure 8.7): the virtual museum tourist wants to book a tour. The reading pattern is as follows: the <system actor> wants to <use case name>.

An official definition of a use case is given by the UML specification [106]: "When a UseCase applies to a subject, it specifies a set of behaviors performed by that subject, which yields an observable result that is of value for Actors or other stakeholders of the subject." That definition defines the model element "UseCase" and does not contain any methodology aspects. They are added for instance by the SYSMOD methodology [145, 147]: the system use case always has at least one actor is started by one trigger from the system context and ends with a result that fits to the intention of the trigger. The behavior between trigger and result is temporally contiguous, that is, there is no temporal interruption provided by the system (temporal cohesion).

The most important requirement for a system use case is the temporal cohesion. The system use case is a specification of a complete collaboration between actors and the system. For instance, the use case "Book a tour" ends with a booked tour. That is why the actor "Virtual Museum Tourist" has triggered the use case. The result satisfies the actor and ends the interaction. A function like "Select available tour" is not a use case. An actor won't use the system just to select an available tour and stops the interaction when the tour is selected. That function is only a part of a more comprehensive use case.

A system actor that triggers or participates in a use case does not need to be a human actor. It is a little bit old fashioned that use cases are mostly described in combination with human actors. Nowadays, systems often communicate more with other systems than with humans. Figure 8.6 shows a use case that is only connected with an external system. We've observed in real projects that new use cases and requirements were identified when applying the use case analysis with nonhuman actors.

To avoid redundancy you can create abstract and secondary use cases. An abstract use case is a general description of a system/actor interaction. It covers the common parts of similar concrete use cases.

The use case "Change customer data" and "Delete customer" in figure 8.5 are similar, but not equal. The common parts of their behavior are specified by the abstract use case "Manage customer data." The two concrete use cases only specify the differences.

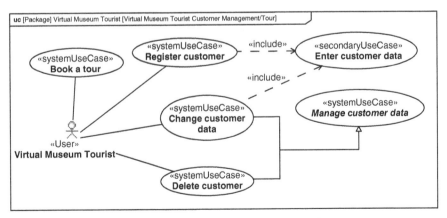

Figure 8.5. System use cases, abstract use cases, secondary use cases.

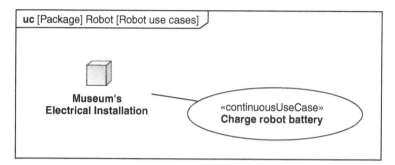

Figure 8.6. Continuous use case "Charge robot battery."

The secondary use case "Enter customer data" specifies a behavior that is used in the use case "Register customer" and "Change customer data." To avoid redundancy, it is specified only once by the secondary use case and included by the include relationships.

The secondary use case is a use case fragment and not a complete use case according to the methodology-based definition of a system use case. For instance, there is no trigger from an actor or the use case provides no results according to the trigger. The concept of a secondary use case is not part of SysML, but commonly used in practice.

Another special use case kind is the continuous use case. It describes a use case that continuously delivers results. The trigger could be an external or an internal event. For instance a state switch. The continuous use case "Charge robot battery" in Figure 8.6 starts when the museum robot is connected with the charger station and stops when it is disconnected. The charger is not an actor of the use case, since it is part of the system.

The use cases are identified by checking each actor of the system context. Most actors trigger a use case or are involved in a use case.

8.4 DESCRIBE USE CASE FLOWS

Use cases are refined by single steps, which we call use case activities or use case steps. We phrase use case activities from the system's perspective, for instance "Get energy" instead of "Inject energy." Note that the use case itself is phrased from the actors perspective as well as the use case activity that represents the whole use case (Figure 8.7).

The control flow specifies the execution order of the use case activities. The object flow specifies the relationship of output objects to input objects of the single use case steps. Figure 8.8 shows the control and

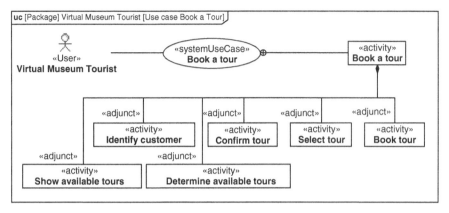

Figure 8.7. Use case activity "Book a tour."

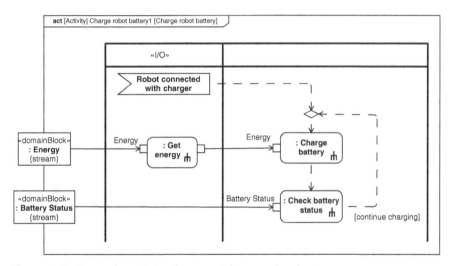

Figure 8.8. Example activity diagram *Charge robot battery.*

object flows of the use case "Charge robot battery" in a SysML activity diagram. The control flow edges are dashed, the object flow edges are solid lines. See Section A.3.2 for a description of SysML activity diagrams.

Each use case step could be further refined with another SysML activity. The refining activity—depicted by another activity diagram—is called by the use case step which is a so-called call behavior action that could be recognized by the fork symbol in the lower right corner of the use case step rectangle.

We recommend to model each use case step with a call behavior action and appropriate activity. The called activity could be empty and just have a name and short description if no further refinement is necessary. That way it is easy to do a refinement later and to use activity trees as another view on the use case activities (Figure 8.7). Activity trees are a valuable view on the system behavior and a supporting tool to create a functional architecture (Chapter 14).

It is a good practice to separate behavior that handles system inputs and outputs from the "real" system behavior. The input and output behavior is more unstable and often closely related to specific technologies than the system behavior. We follow the principle to separate unstable from stable parts (Section 7.5) and separate the use case steps accordingly. We assign them to separate I/O activity partitions as depicted in Figure 8.8. It is also an important preparatory work for the FAS method (Chapter 14).

A more detailed description of modeling use cases with activities is provided, for example, by Weilkiens [145, 147].

8.5 MODEL THE DOMAIN KNOWLEDGE

The domain knowledge represents entities from the systems domain that are known and used by the system. You find the entities in the object flow of the use case activities. The domain entities are the types of the inputs and outputs of the use case activities. The domain knowledge is modeled in SysML with blocks and associations and depicted in a block definition diagram. It is often also called simply domain model.

Figure 8.9 shows an extract of the Virtual Museum Tour (VMT) system domain knowledge. The blocks have the stereotype «domainBlock» from the SYSMOD profile [145, 147] to explicitly mark them as blocks of the domain knowledge. Each domain block could have value properties. Associations between domain blocks specify reference properties that have other domain blocks as its type.

For instance, the association end *tour* specifies that at any time a "Tour Booking" instance is always connected with exactly one "Tour" instance. "tour:Tour[1]" is a reference property of the domain block "Tour Booking."

In software-intensive parts of a system the appropriate domain blocks are closely related to the conceptual data model. However, the domain knowledge does not follow the design rules of a real database model. In addition, the domain knowledge contains also domain blocks that

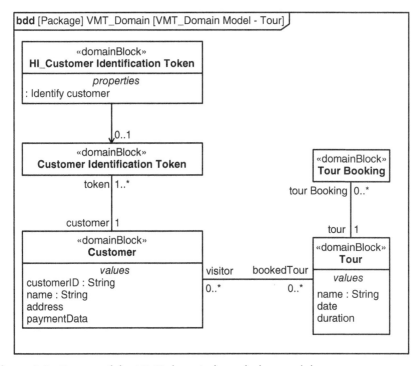

Figure 8.9. Extract of the VMT domain knowledge model.

represent real physical things and no data. For instance starlight in a telescope model [54] or "Energy" in Figure 8.8.

The domain blocks are the types of the objects that flow through the use case activities. The separation of the input and output behavior leads to a separate layer in the domain knowledge. In Figure 8.9 the domain block "HI_CustomerIdentificationToken" is the representation of a customer identification token in the context interface layer. See Section 9.5 about layers.

Another part of the domain knowledge is the list of units used in the system model. Most units are common and could ideally be retrieved from a model library. For instance the "ISO 80000" library (Section A.6). New units that are defined for the system of interest are part of the domain knowledge and good candidates to be extracted into a separate model library for reuse in other projects of your organization.

Chapter **9**

Perspectives, Viewpoints and Views in System Architecture*

9.1 OVERVIEW

Try to show more boxes and arrows in one single diagram than one can easily understand, then you may be able to reproduce a problem that P.B. Kruchten describes in the beginning of his article "The 4 + 1 View Model of Architecture" [81]. He addresses the problem of architectural representation that overemphasizes certain aspects of development while neglecting others. Kruchten's "4 + 1 View Model" was developed as a remedy to the problem. It is based on the approach of addressing different stakeholders' concerns with different views. The "4 + 1" views are the logical view, the development view, the process view, the physical view and the scenarios,[1] which tie the elements of the other four views together. Kruchten's work was targeted at software architecture. Nevertheless, the notion of separating different stakeholders' concerns via different views also holds for system architecture, as we have seen in Chapter 6.

The term *view* is known from Section 7.7, which is about the separation of view and model. A view in that sense is a general means of

---

* Together with Matthias Dänzer, Bernafon AG.
[1] In the context of this book we would prefer the term *use cases* on *scenarios*.

Model-Based System Architecture, First Edition.
Tim Weilkiens, Jesko G. Lamm, Stephan Roth, and Markus Walker.
© 2016 John Wiley & Sons, Inc. Published 2016 by John Wiley & Sons, Inc.

representing only as much of the available information as is practical. To distinguish it from views that offer specific details from the system architecture for addressing a given stakeholder's concern, we call the latter ones *architecture views*. Their role in architecture descriptions has been discussed in Chapter 6.

When looking for suitable stakeholder-specific views, we may find a set of different stakeholders who are all interested in the same kind of information about the system, but still have different focus areas. For example, both logistics people and maintenance people may be interested in the geographical spread of the system, but while logistics people may be more interested in the weight and the maximum dimensions of objects to ship to different locations, the maintenance people may be mainly interested in the spare parts needed at different local stocks. Both kinds of stakeholders thus need different views, but they are both interested in the geographical kind of information.

In order to group information of the same kind, we have adopted the notion of *perspectives* from the TRAK architecture framework that will be explained in Section 16.3.7. We can use perspectives as a means of grouping or categorizing architecture views, as we have seen in Chapter 6. The perspective from which both logistics and maintenance people in our example would like to see the system is one whose focus is on geographical deployment of the system, but as explained earlier, each of them would like to see a different view of the system.

Since our definition of views is tied to addressing stakeholders' concerns, we would like to be able to serve stakeholders with accurately filtered information. Therefore, we have chosen to reserve the notion of views for means of presenting narrow subsets of the information that one would traditionally expect in something like "the physical view" of the system. In a model-based approach, it is easy to maintain larger sets of such narrow views, because the model ensures their consistency.

Having the possibility to address stakeholders concerns in a very focused way with these proposed narrow views, we can exclude the existence of something like "the physical view," because we have so far not encountered any stakeholder who likes to see the complete modeled information about all physical elements in the system at once. We have therefore used the notion of perspectives for covering aspects like the "physical" representation of the system. This gives us different perspectives than the ones proposed in TRAK. While these are Enterprise, Concept, Procurement, Solution, and Management, the perspectives we propose are inspired by typical categorizations in the literature like "functional," "physical" (e.g., Refs [79, 139]), "behavior" (e.g., Ref. [53]), "layered" (e.g., Ref. [91]) or "deployment" (e.g., Ref. [123]).

This gives us the *functional perspective*, the *physical perspective*, the *behavioral perspective*, the *layered perspective*, and the *system deployment perspective*. The latter one is derived from the deployment views of software architecting but has been extended for the sake of systems architecting. It will allow to accommodate for environmental and spatial information but also for the geographical spread of the system that was mentioned in the above logistics and maintenance example. We have dedicated a section in this chapter to each of the perspectives. For example, Section 9.6 will provide more details about the system deployment perspective. At the end of the chapter, we will discuss how the different perspectives relate to each other and to other concepts we have discussed in this book. Each perspective can have its own domain knowledge model according to Section 8.5. We will see in Section 9.5 that the domain knowledge model of the layered perspective can be even subdivided by layers.

9.2   THE FUNCTIONAL PERSPECTIVE

Seen from the functional perspective, the system is considered as a set of interrelated functions, that is of different relationships between inputs and outputs of the system itself or elements within it (see also Chapter 14).

A very simple and abstract representation of functions is a hierarchical decomposition of system functionality into functions and their subfunctions. Howard Eisner presents a simple informal diagram that can express such a decomposition ([33], p. 146). Figure 9.1 shows how parts of the Virtual Museum Tour system could be functionally decomposed in such a nonformal diagram. The elements shown in the figure are called functional blocks. They could also be modeled more formally in SysML. This is described later.

Figure 9.1 illustrates an important notion of the functional perspective as it shall be here: the functional blocks are static. This means that functions are represented independent from preconditions needed for making use of them or constraints on the sequence in which they can be provided by the system. These latter aspects would belong to the behavioral perspective (Section 9.4). Therefore, functional blocks are very similar to "system elements" according to Chapter 4 and their interrelations can be seen as system element interactions. In consequence,

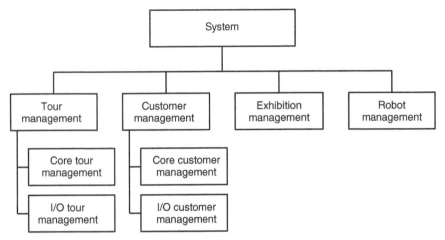

Figure 9.1. Excerpt from the functional decomposition of the Virtual Museum tour system, presented in an informal way.

interrelations between functions can be modeled by the same means as interfaces between system elements. This will be discussed further in the context of the so-called FAS method in Chapter 14.

The specialty of functional blocks is that they may not be system elements of the physical system. For example, the functional group "Tour Management" of the Virtual Museum Tour system may be scattered across different servers and client computers or handheld devices in the museum itself, in the cloud and within the virtual museum tourist's reach. People working with views from the functional perspective may need abstract thinking, due to the possible difference between the system as it is seen from the functional perspective and the real system as it is seen when disassembled.

The functional perspective allows for describing the complete system conceptually, even if some implementation details are not yet known. It can provide a high-level overview of the system that does not only enable an understanding of the system's principles of operation, but can also be used to trace fulfillment of functional requirements long before the implementation details are settled. The latter can be exploited to create rough work breakdown structures for covering functional requirements in a very early phase of development. Views of the functional perspective are thus not only interesting for technical stakeholders but potentially for development managers or project leaders as well. Users of the functional perspective should however pay attention to the fact that solutions and constraints necessary for satisfying nonfunctional requirements are often underrepresented or even not visible via the functional perspective.

It may be hard to see the value of the functional perspective for people concerned with very detailed engineering problems because it is very abstract. This is one reason why we have seen organizations fail to introduce the functional perspective. Indeed it is more intuitive for people with engineering background to model the system from a physical perspective. In cases in which the functional perspective is omitted, we do however see the risk of making models too detailed, resulting in over-modeling or redundancy between the architecture description and the development documentation produced outside the scope of the system architect.

On the one hand, the functional perspective is thus beneficial for keeping the overview and ensuring an appropriate abstraction level in architecture descriptions; on the other hand, it may be hard to understand or at least hard to accept for parts of the organizations. This is why we recommend to system architects who consider the functional perspective to be valuable to their organization that they declare this perspective as one of the most important ones. Of course, this is dangerous, for example, regarding the potential underrepresentation of nonfunctional aspect that was already mentioned. Still, more physical and behavioral perspectives can usually be introduced very intuitively and can just be made without too much need for justification, whereas we have made the experience that the functional perspective needs through explanation and should thus be in the focus of the system architect until it is well established.

9.2.1 SysML Modeling of Functional Blocks

To reflect the static nature of functional blocks, we recommend modeling them as blocks in SysML. This way of modeling has been proposed by Lamm and Weilkiens [83, 84] and also by Fernández-Sánchez et al. [38]. The corresponding representation of Figure 9.1 is shown in Figure 9.2 in a SysML block definition diagram.

The flows of information (signals, data), materials, force or energy, between different functions can be modeled via ports and connectors. Chapter 14 elaborates on this way of modeling in the context of the so-called FAS method.

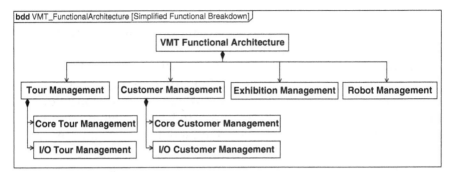

Figure 9.2. Excerpt from the functional decomposition of the Virtual Museum Tour system, in SysML.

9.2.2 Views for the System Architect

The system architect will need certain views for the own work on the system architecture, for example:

- A set of views showing only excerpts from the functional architecture, for example, around a certain feature of the system. One can create a set of them which together covers all functional elements. They can be used, for example, to make a rough assessment on the status of the work on different features of the system.
- A view showing all functional blocks in the system down to a certain hierarchy level. A typical concern of the system architect to be assessed with such a view is an impact analysis that is triggered by a change request. One option could be to generate such views only if needed and to discard them after being used, in order not to have too many details to maintain. As an example, let's assume that the Virtual Museum Tour System would need to be changed in order to comply with a new privacy protection policy of a certain museum that requests computers storing customer data to be locked up in closed rooms. Of course not all components of the system can be locked up in a closed room. A typical procedure for the impact analysis would be as follows:
  1. Generate a list of all functions in the system.
  2. Walk through the list systematically and highlight all functions that process customer data.
  3. Assess which subsystems provide the highlighted functions and find out if a repartitioning is needed to subsystems that can be locked up in a closed room.

4. Assess the effort and risk that is related to the repartitioning task.

There are many more views that the system architect might like to create either in an ad-hoc manner or well planned in advance. The earlier examples are those that we have observed maybe most often.

9.2.3   Different Views for the Stakeholders of Different Functions

At the beginning of this chapter, we explained that stakeholders who are interested in the same perspective of the system may still like to have different views from the same perspective. Model-based system architecture enables the creation and maintenance of very focused views for different stakeholders.

The functional perspective can be used to scope development activities that are focused on one function of the system (e.g., Ref. [82]). Teams performing such activities might like to see their own view of the system from the functional perspective. This is typically a view on the function in focus of the current activity, together with those functions it has interrelations with. In that case, the system architect can create a view that is constructed around the functional block of interest.

Let's assume that a certain team in the development of the Virtual Museum Tour System has the task to implement the tracking of the museums different robots' utilization, so for example if the robots are available and at what position they are. Such a team could have a functional block "Robot Utilization Tracker" as its scope, but it would also like to know how to obtain position information and which other functions in the system would need the utilization information. Figure 9.3 shows an example view for such a team: a SysML internal block diagram shows excerpts from the functional architecture. The functional block of interest is shown larger than the other blocks on the same hierarchy level, and the direct interrelations to other functional blocks are shown. Blocks without such direct interrelations to the block of interest are omitted in the diagram.

In the given example, it is possible to make a formal definition of the contents to be displayed in such a view, like "the block of interest and the blocks that have interfaces via ports with it." If a high number of views according to the same definition needs to be maintained for different functional blocks of interest, then automation features of the modeling tool can be helpful: an automatic view assessment tool can be created.

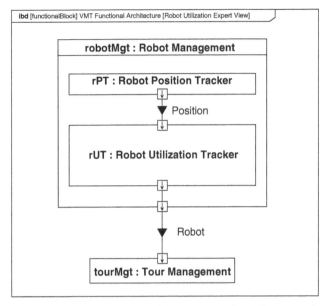

Figure 9.3. A view from the functional perspective for a functional team working on utilization of robots.

It checks whether the views that should comply to the definition indeed show all model elements that the definition requires to be shown.

9.3 THE PHYSICAL PERSPECTIVE

Seen from the physical perspective, the system consists of system elements according to Chapter 4. Interface control documents (e.g., Ref. [79]) may also be assigned to the physical perspective.

One can look at multiple levels of abstraction in the physical perspective. We saw in Section 7.1 that System Architecture may consist of the base architecture level, the logical architecture level, and the product architecture level. All three levels can have a physical perspective.

9.3.1 Logical Architecture Example

The logical architecture is a representation of the system in which the system elements are defined according to the technical concept for implementing the system but are not defined concretely enough for implementing the system. This will be illustrated here, using an

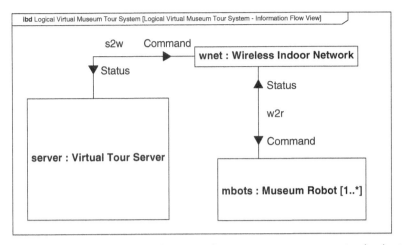

Figure 9.4. Example view on the Virtual Museum Tour system in the logical architecture.

example. See Chapter 14 for the general definition of the term "logical architecture."

Figure 9.4 shows an example related to the Virtual Museum Tour system: The "Wireless Indoor Network" is a solution for the communication inside the museum building. The technical concept is to use a wireless network solution and for example not a proprietary point-to-point radio protocol for communicating with the robots. The logical architecture does not define which wireless network standard the network will comply to and how the routing of information inside the building will happen.

Interfaces can be defined roughly in the logical architecture. Figure 9.5 specifies a high-level specification of the interfaces from Figure 9.4 by defining a domain knowledge model of the logical perspective in a SysML block definition diagram.

9.3.2 Product Architecture Example

The product architecture is concerned with the concrete technical implementation of the solution (see also the definition in Chapter 14). To continue the example from the previous section, the product architecture is thus about the concrete network technology used to establish indoor communication between the server in the museum and the robots.

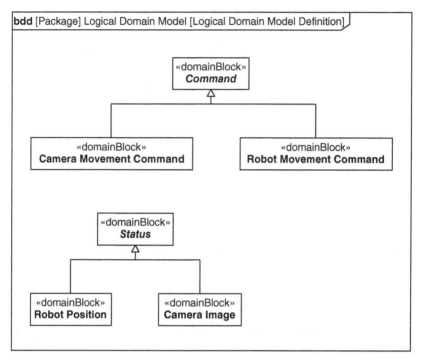

Figure 9.5. An interface view in the logical architecture: details on interfaces are defined by specifying the data flows according to Figure 9.4 in a domain knowledge model whose scope is the logical architecture.

Let's consider that the wireless indoor network is established by placing wireless access points that provide a wireless network in the museum and connects to the museum tour server via a powerline network,[2] in order to save cost for the installation of a wired network connecting them with the server. This means that both the server and the wireless access points need connections to the museum's electrical network in order to be provided with both power supply and access to the indoor network.

---

[2] Powerline networks use existing electrical networks in order to transmit data packages. The user can plug them into existing electrical outlets. They will contact each other via a high-frequency carrier signal transmitted on the electrical power network and establish a local area network for computers or similar devices. Here we choose this technology to illustrate that views are nonorthogonal: the electrical and communication aspects of the system overlap due to communicating via the electrical network. The consequences of nonorthogonal views will be elaborated on in Section 12.1.3.

Only the physical perspective will show the concrete solution "Powerline Network," while the logical perspective has just shown communication flows, as we saw in Figure 9.4. Some stakeholders may also like to see communication flows from the physical perspective. They can be provided with a view like the one in Figure 9.6, where the server and the network access points and some of their connections are shown, with a focus on those connections needed for communication only.

A stakeholder who has to care about installation procedures and installation cost of the system may not be interested in the communication inside the system but rather in the number of mechanical and electrical connections. This kind of stakeholder may be provided with a view very similar to the previously discussed one, but now with an additional focus on the mechanical aspects of connections. An example is shown in Figure 9.7 in a SysML internal block diagram. Note that in comparison to Figure 9.6 the wireless connection with the robot is no longer shown, because it is neither an electrical connection nor a mechanical connection. Later we will see how the system deployment perspective may provide additional facts that are relevant for the installation of the system in terms of wireless network performance (Section 9.6).

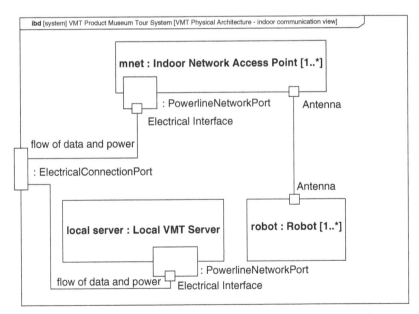

Figure 9.6. Example view on the Virtual Museum Tour system from the physical perspective with focus on communication.

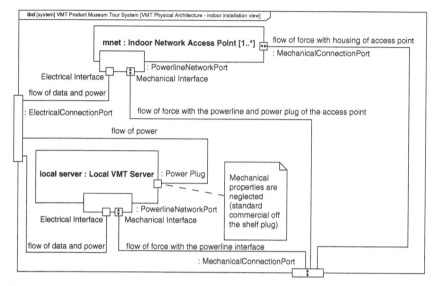

Figure 9.7. Example view on the Virtual Museum Tour system from the physical perspective with focus on installation.

9.4 THE BEHAVIORAL PERSPECTIVE

The behavioral perspective defines what the system shall do as a sequence of actions over time. It is concerned with state transitions in the system and the timing and sequence of function calls or information exchange via interfaces. SysML state machine diagrams, sequence diagrams and activity diagrams with control flows usually show views from the behavioral perspective. The appendix contains examples for such diagrams.

Particularly, software engineers often ask for behavioral specifications of the system, because state machines and programmed sequences of events are usually realized via software or embedded software inside the system.

9.5 THE LAYERED PERSPECTIVE

9.5.1 The Layered Approach

The layered approach originates from software architecting, but as we will explain there are good reasons to apply it to systems architecting as well. After a short discussion of layers in software architecture, we will generalize the idea of layered abstractions to systems architecting for systems with different information processing components.

The layered approach in software architecting is based on separation of concerns. It can be seen as a modern form of a modularization principle formulated by Parnas [114], stating that modularization should aim at defining modules in a way that each module hides a difficult design decision or a design decision that is likely to change. The benefit is that interfaces with other modules do not need to change if a design decision about the innards of a module changes. In other words, the layered approach results in modules with low coupling (see also Section 7.3).

In a layered approach, the modules are assumed to be "stacked" in a certain order and are therefore called *layers*. Indeed a set of several connected layers can be called a "Layer Stack." The notion is that a layer can only have an interface with an adjacent layer such that information has to be systematically transformed to levels of the stack one by one instead of being able to bypass parts of it. A typical advantage of such an approach is to avoid uncontrolled dependencies between different parts of the system and to enable changing parts of the solution without impact on other areas of the system. This enables flexibility in choosing and changing the concrete technological solution for the implementation. In cases in which flexibility is a dominant quality criterion, the different layers can be chosen to represent different abstractions of implementation details. In that case, layers are stacked in order of ascending abstraction level.

One example of a layer that both hides a design decision and enables the flexible exchange of a solution is a "hardware abstraction layer" in a software product. It is designed to offer a standardized interface for using hardware toward the adjacent layer that is higher up in the layer stack. This means that the upper layers of the stack can use hardware, independent of its concrete realization. The concrete information exchange with the hardware is hidden from the upper layers. This enables porting of the software product to a completely different hardware without changing anything in the implementation of the upper layers. Only the hardware abstraction layer needs to be adapted to the new hardware, but it will still offer the same interface toward the upper layers. In a variation of this approach, the software product can dynamically create different instances of the hardware abstraction layer, each being able to handle a different hardware solution. Dänzer et al. [31] have provided an example from the hearing care field, where a layered approach was used in a software product that interoperates with diagnostic devices from different manufacturers. The layered approach was chosen with the aim to enable an easy integration of the different manufacturers' devices.

All the previously said can be handled by software architects, as long as the layered approach is used inside one software product or inside one software subsystem of a system. So why should systems architecting be concerned with layered architectures? That is the topic of the next chapter.

9.5.2 The Layered Perspective in System Architecture

An early example of a layered approach that can be applied to partially nonsoftware systems is the OSI reference model according to ISO/IEC 7498-1:1984 [62], which has been replaced by a new revision in the year 1994. It defines a standardized way of interconnecting systems, based on the notion that higher layers can be specified independent of the technology used in lower layers. Two applications on different devices can exchange information, without a dependency on the carrier technology, so for example no matter whether the information is transferred via a cable, a fiber optic connection, or a radio transmission. The lowest layer of the OSI reference model is concerned with the physical transmission technology, which would imply the use of electronic or optoelectronic hardware in the given examples. The OSI reference model thus enables the interconnection of devices with considerable nonsoftware technology into a system, with the aim of interconnecting different nodes inside the system, which can exchange information without being developed for a special information interchange medium.

This leads us to one advantage of the layered approach for systems architecting: if the whole system is based on one shared layer model, then different entities inside the system can exchange information without being sensitive to changes of the solution in other areas of the system. An example based on the Virtual Museum Tour system may underline this: Assume that a user of the system has opened the application for controlling the museum robot on a mobile device, in parallel with a geo information system that can display satellite pictures and location information based on the longitude and latitude of a position on earth. The user may now like to see where on the satellite picture the robot's current location is. If the robot position can be retrieved in global coordinates, then this is possible. It may thus seem practical to handle robot locations as global position information as soon as the information needs to be exchanged with other devices or needs to be presented to the user. On the other hand, the controllers that process

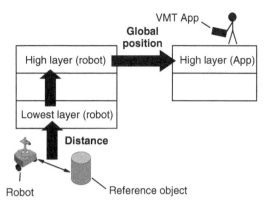

Figure 9.8. An informal view in the layered perspective.

the information from distance sensors on the robot may not have all the information necessary to convert measured distances into global positions. When organizing the Virtual Museum Tour System according to a layered approach, a higher layer in the layer stack might thus represent robot positions in global coordinates, whereas a lower layer may perceive position information as a distance value. This is shown in Figure 9.8, with a strong simplification: It is not shown that the communication between the robot and the VMT App needs some calls to lower layers, in order to use a communication service. A view without this simplification will be provided in Section 9.5.5.

If a whole system has to be designed according to a common layered approach, then it is not sufficient to define the layered architecture as a part of the software architecture. Heterogeneous systems may be composed of very different kinds of information processing subsystems, involving different information processing components with different architectures. Still it may be desirable to impose a system-wide layer model on all the subsystems. This has to be done via an artifact that is visible across the different stakeholders involved in creating the different subsystems, so best by the system architecture description. The *layered perspective* will be the one with the views needed to impose a common layer model on the system and make it visible to different stakeholders.

The layered perspective has one fundamental difference from the functional and the physical one: while the latter ones are based on a decomposition approach, the layered perspective is based on a service approach. The physical perspective for example allows us to show how a system is decomposed into subsystems, which are part of the system. In a layered approach, a lower layer is not part of a higher layer, it rather offers a service to it: Maier [91] states that layers have an „is-used-by" relationship as opposed to the „is-a-part-of" relationship in decomposition approaches. Maier also points out the challenges in handling layered approaches in systems architecting, which is traditionally based on decomposition approaches.

A systems architecting approach based on layered architectures should thus be considered a challenging one, and to our knowledge there has until recently not been much interest in using a layered perspective like we propose it in system architecting. This is why we would like to keep the challenges of the approach manageable and therefore define that the scope of layered perspectives is limited to information processing aspects of the system. As a consequence, an important aspect in layered architecting will be to translate information flows from the domain knowledge model (Section 8.5) into real-world representation of the information. This will be discussed further in Section 9.5.3. As a rule of thumb, this means that system elements which are not needed for handling information objects from the domain knowledge model are very likely to be out of scope in the layered perspective.

A central notion of layered architectures is the *information hiding*, this means: information that may be present on one layer may not be available on another layer. As a consequence, information that is highly dependent on the solution or design decisions in certain areas of the system cannot accidentally be used in other areas of the system, which would lead to side effects and undesirable bad dependencies. In the museum robot example, the exchange of the distance sensor technology may lead to a different representation of distances. The technology-dependent representation should be encapsulated inside the driver layer, such that it is hidden from other layers. As long as only an isolated part inside that layer processes the sensor output, it is easy to analyze the impact of a changed sensor technology: only the part of the layer that uses the sensor output needs to be redesigned (see also the pattern "separate stable from unstable parts" from Section 7.5). The interfaces to other layers will remain the same, and in consequence only the driver layer needs to be updated. In a solution without information hiding, the output of the distance sensor can in theory have been used by a developer working on an area of the system whose relation to the distance sensor is not directly obvious. This comes with the risk of forgetting this area of the system when analyzing the impact of a changed sensor technology. Information hiding can thus facilitate change management and maintenance of the system architecture.

9.5.3 Relation to the Domain Knowledge Model

The information hiding can imply that objects from the domain knowledge model (abbreviation from here on: *domain model*) may not be available on certain layers. We have already discussed the example of the robot position, which has different representations on different layers. To model the different representation of the same kind of information on different layers we propose to extend the domain model with layer-specific domain objects. While the domain object was so far an artifact of requirements engineering, the proposed additional blocks are artifacts of systems architecting. These can then be traced to the corresponding domain object of the original domain model from requirements and use case analysis. Figure 9.9 shows this, based on the mentioned example: The "Position" from the original domain model represents some kind of position information to be processed by the system, based on findings during requirements analysis that the system needs to process positions. The layer-specific representations of position information are as follows:

- "Global Position": the kind of position information to be used on the application layer and to also be displayed to the users of the system
- "Distance": the kind of position-related information to be used in system elements close to the distance sensors
- "Museum Internal Position": the position compared to one reference point in the museum, to be used on an intermediate layer.

The blocks on the right-hand side represents the different layers. In the shown example, they have been chosen with the aim of separating different abstraction levels. Their modeling will be further discussed in Section 9.5.5. The «trace» relationships from the domain blocks to the «layer» blocks assigns each domain block to a layer. The dependency relationships that point diagonally up to the right indicate that the domain blocks on a given abstraction level are also used on the higher adjacent layer. This is because the information is handed up through the layer stack and needs to be transformed before being in the proprietary information of the higher layer. Systems architecting will typically own every «domainBlock» object that is traced from a «layer» object,

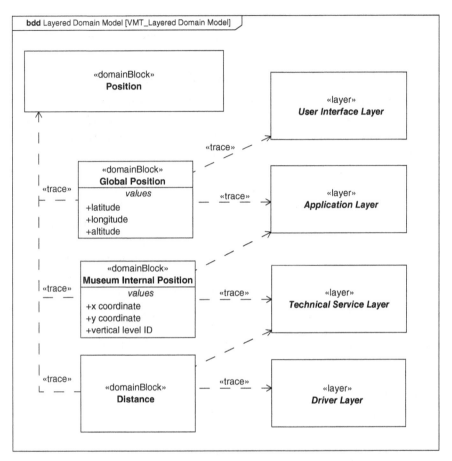

Figure 9.9. Layer-independent domain block "Position" and the derived layer-specific domain blocks.

whereas layer-independent blocks like "Position" in Figure 9.9 are typically artifacts from the requirements and use case analysis.

9.5.4   Architecting the Layers

When defining a layered architecture, one needs criteria on how to separate different aspects into different layers. Here are some considerations:

- We already saw that Parnas [114] recommends hiding difficult design decisions or those that are likely to change. The latter is related to the pattern "separate stable from unstable parts" from Section 7.5.

- It should be possible to release layers separately. This can save reverification effort after changes because the similarity with a prior version of the system is given for those layers that were not updated during the change.
- Layers should be separately testable (at least when combined from the bottom up). System architects together with the appropriate verification stakeholders can assess whether the following scenarios are easily feasible with the chosen layered architecture, in case these scenarios are desirable:
  - Hardware together with its driver layer can be tested by means of test software running on top of the driver layer, without the need for a human user to operate the system. This enables hardware testing before the completion of software.
  - Conversely, the lower layers can be replaced by a hardware emulator, enabling testing of user interfaces or application logic even before the hardware is available.
- It should be possible to generate test inputs for domain objects that are handed over via interfaces between layers.
- Assess whether it is practical to perform a test with the given domain objects as an input.

9.5.5 SysML Modeling of Layers

We will explain an approach for modeling layers in SysML. It has been based on a proposal by Dänzer et al. [30]. We propose modeling abstract blocks to represent layers. Concrete blocks are used for the part properties of the layered architecture. They can be modeled as specializations of the abstract blocks, which makes them inherit, for example, the ports from those.

Figure 9.10 shows the definition of layers for the Virtual Museum Tour system. On the left-hand side, the composition of different layered parts into a whole is modeled. The picture shows an incomplete example with just the blocks necessary for explaining the modeling approach. On the right-hand side of the figure, the different layers are modeled as abstract blocks with the stereotype «layer». Blocks can be assigned to the given layer by means of a generalization relationship. Layered elements which can be implemented in a physical subsystem are modeled as concrete blocks with the stereotype «layer». In the figure, we see that there are two blocks of the "Driver Layer" kind: the "Robot Driver Layer - Camera" and the "Robot Driver Layer - Distance Sensor." Via

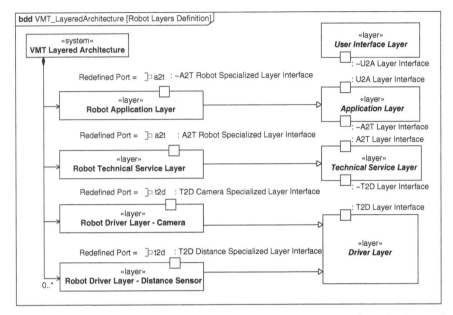

Figure 9.10. Example view on the Virtual Museum Tour system from the layered perspective.

the generalization relationships, the concrete blocks inherit all ports from the abstract blocks on the right-hand side. This cannot be seen in the figure. However, one can see that some of the inherited ports are redefined into more specialized ports. The approach of modeling ports for the whole abstraction level and then redefining them where appropriate ensures that there is a standardized way of interfacing layers to each other, which is important, for example, to maintain the benefit of keeping control of side effects. The naming of the ports will be discussed later. The diagram shows only few blocks, but in the layered architecture we present in the following, each concrete «layer» block is derived from exactly one of the abstract blocks. This ensures that the different subsystems in the system adhere to the system-wide valid layer split. Here, the different layers have been chosen to separate different abstraction levels. This means that all parts of the system with the given layers are based on a common way of abstraction and can therefore easily be interconnected. In analogy to this, the developers of the different parts of the system have a common basis for structuring their work products, which can make it easier for them to understand each other's area of development and communicate with each other.

Figure 9.11 shows how the ports for information exchange between adjacent layers are defined: in general, they are named according to the layer boundary on which they reside. For example, ports residing on the boundary between the user interface layer and the application layer will be identified using the abbreviation "u2a."

Depending on the abstraction level, there are different operations in relation to the robot position. They are directly related to the domain model we saw earlier in Figure 9.9: while something of type "Distance" from the domain model is returned from a "GetDistance()" operation on an interface that is provided by the Driver Layer, something of type "Global Position" is returned from the corresponding operation on an interface that is provided by the application layer. Information hiding is in place and ensures increasing abstraction level from bottom to top of Figures 9.9 and 9.10. It should be noted though that the vertical order of blocks in the diagrams has been made arbitrarily during diagram layout. It is not part of the model but only expressed in the given diagram.

In the model, the order of layers can be defined by means of their interconnections in the internal block diagram. An example is shown in Figure 9.12: We see parts of the layer stacks of the server application and of the robot in the museum. Again some details are omitted. For example, it is very likely that a server application has a user interface layer, but within the view according to Figure 9.12 this is not shown. The view has been created to show a SysML version of parts of the informal drawing in Figure 9.8 that we saw earlier when the notion of information hiding was exemplified with the conversion of distance information into global position information. In that informal figure the communication between the robot and the user's VMT App was not shown in detail. Figure 9.12 now shows this detail: a "Local Network Driver Layer" on both the server side and the robot ensures that the server and the robot can exchange information via a local network. The network connection itself is out of scope, seen from the layered perspective. It would belong to the physical perspective. In order to still enable the reader of the diagram to follow the information flow, a connection has been modeled between "LAN Server: Local Network Driver Layer" and "LAN Robot: Local Network Driver Layer." This connection has no precise technical meaning in the context of the layered perspective, but since views are created for the stakeholder, the creation of clarity can also give meaning to a model element. In the diagrams shown here, all connections that connect to layers without ports are such cases in which an information flow has been visualized in an informal way. The connections representing context interactions by contrast have a technical meaning.

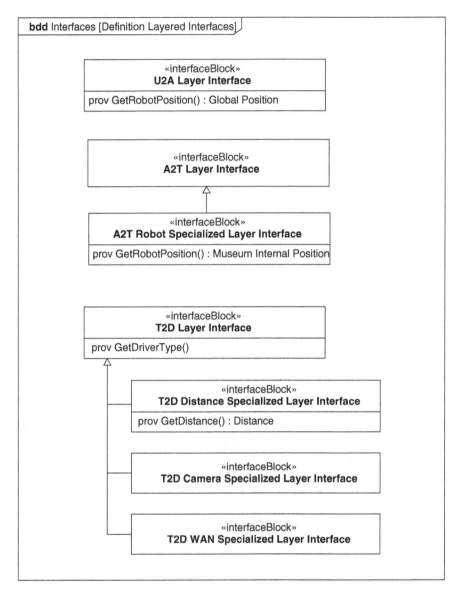

Figure 9.11. An interface view from the layered perspective: details on interfaces are defined by specifying the port types of ports from Figure 9.10 in a block definition diagram.

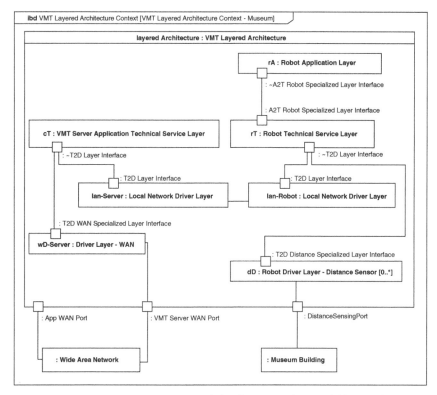

Figure 9.12. SysML representation of the layer stacks on different devices, as sketched in Figure 9.8.

They represent information exchange between the system and external elements. In this case, the "Wide Area Network" is one of them. We see that the server can access it via the corresponding driver layer. The Wide Area Network (WAN) is outside the system, but it transmits information to another element inside the system, that is, the VMT App. Figure 9.12 only shows a second port that is connected with the WAN, but the figure would have been overloaded had we shown the layers of the VMT App as well. Therefore, a second Figure 9.13 has been created to show that aspect.

When using the modeling approach according to [30] with SysML 1.3 or 1.4, the interfaces between ports of «layer» blocks should be proxy ports (see SysML reference in the appendix), because layers can only offer operations that they own internally. In the figures shown here, we have omitted the «proxy» stereotype to avoid overloading them completely.

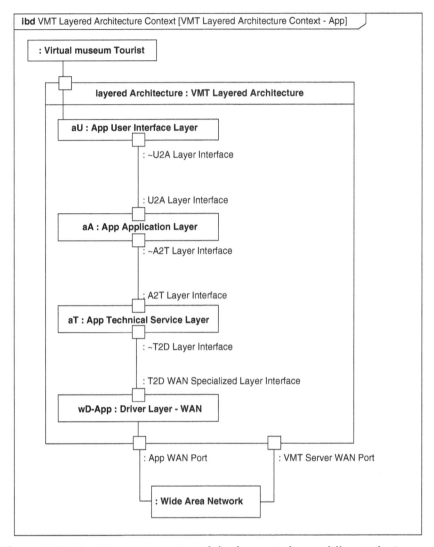

Figure 9.13. SysML representation of the layer stacks on different devices, as sketched in Figure 9.8.

9.6 SYSTEM DEPLOYMENT PERSPECTIVE

In software architecting, deployment describes how software artifacts are distributed across the nodes of the runtime environment. These software artifacts could be the outcomes of the software build process, configuration files, execution environments like virtual machines or application servers, libraries, and frameworks. In the case of an embedded software for a multiprocessor device, a view of the deployment perspective could describe how the different software modules are distributed to different processor cores. How can this perspective be generalized to be applicable for systems architecting? Kossiakoff and Sweet [79] see deployment mainly as the transportation to the operational site and the subsequent installation. In analogy to software architecture, we would like the deployment perspective to be more static in the sense that it describes the system after it has been deployed. Figure 9.14 shows an example of a view of the Virtual Museum Tour system in our deployment perspective, which we call the *system deployment perspective* from now on to make a clear distinction from software architecture via the naming.

Considerations like transportation can still be derived in views of the system deployment perspective, because a picture of the deployed system should allow considerations about how to deploy it. In the example in Figure 9.14, one could for example immediately see that a robot needs to be transported to a museum and lifted to a certain floor. The logistics people from the example in the beginning of this chapter would like to see the system from this perspective in a view as shown in Figure 9.14.

In their book "Software Systems Architecting" [123], Rozanski and Woods emphasize the aspect of the system's environment in the deployment context, which of course is a runtime environment in the software field. Playing with the word "environment" actually allows us to see that there is a notion of a different kind of environment attached to the *system deployment perspective* like it is exemplified in Figure 9.14: the figure shows that the robot is used in an indoor environment, such

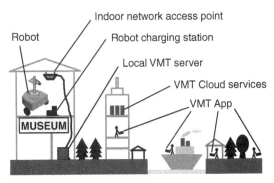

Figure 9.14. A view of the system deployment perspective in informal representation.

that environmental conditions like rain, wind, or fog do not need to be taken into account when designing it.

We also see that location information is contained in Figure 9.14: it is clear that the distance between the installation in the museum and the rest of the system can be considerably large. One can deduce that response times in communication between the "VMT App" of the virtual museum tourist and the robot may become an issue. In the context of software architecting, Rozanski and Woods [123] define a "location perspective" to which they attach these kind of distance effects that result from the absolute location of system elements. In our case, we see no necessity to define this additional perspective, because our generalization of the deployment aspect from software architecting into a system deployment perspective for systems architecting produced location information available within the system deployment perspective.

The system deployment perspective contains elements whose home is in other perspectives. For example, the system elements shown in Figure 9.14 belong to the physical perspective. But it is also possible to make a system deployment perspective with functional elements or layers or even combinations. For example, one can use the system deployment perspective to assess which functions of the Virtual Museum Tour system will be deployed to system elements that are always present in the museum itself. This will allow to consider for example whether a service technician needs to bring a mobile device with an app to service robots in a museum. As a consequence, one can determine if the user interface layer needs to be deployed to the local VMT server in the museum or only to the mobile device of the service technician.

The system deployment perspective can also facilitate interface specifications. In the case of the Virtual Museum Tour system, the architect may like to know if the power connection of the indoor network access point should be at the top or at the side of the device. Via the appropriate view of the system deployment perspective, one can answer this question. In the example in Figure 9.14 the architect would conclude that the power connection is best foreseen at the top of the indoor network access point. Liang and Paredis [88] point out that a position property is a possible part of a port specification. In the example of the indoor network access point, such a property could be used to model the position of the power connection. It would become visible via a view of the system deployment perspective.

Last but not least, the system deployment perspective can be chosen for defining constraints on positions or distances. For example, there is a maximum transmission range of network access points in the

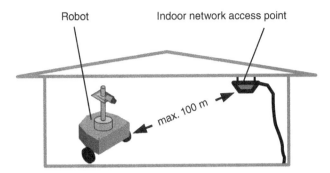

Figure 9.15. A constraint in the system deployment perspective in informal representation.

museum. Figure 9.15 shows an example of expressing the corresponding constraint via the system deployment perspective.

In summary, the system deployment perspectives can provide stakeholders with

- Location-related information
- Spatial and distance-related information
- Environment-related information that specifies in which locations certain system elements are exposed to the environmental effects identified during work on the system context (see Section 8.2) and which of the system elements are affected.

Spatial information of the system deployment perspective cannot yet be modeled in SysML very intuitively. Positions can however be modeled as properties of blocks and constraints on positions can be modeled as constraints in parametric diagrams. As an example, Figure 9.16 shows a SysML representation of the constraint from Figure 9.15 in a parametric diagram.

9.7 OTHER PERSPECTIVES

Of course this book cannot present an exhaustive set of perspectives. There are as many perspectives as there are kind of information to process about the system of interest. The system architect will thus have to define own perspectives in case the ones given in this book are not suitable.

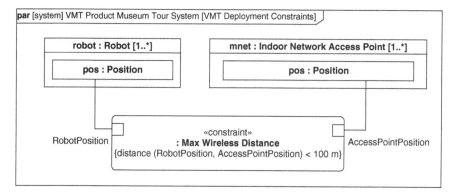

Figure 9.16. The constraint from Figure 9.15 in the system deployment perspective in SysML representation.

There are two perspectives that are too much overlapping with the system deployment perspective to deserve an own section in this chapter, but sill they shall briefly be mentioned:

The *geometry perspective* is one from which geometry and shape are in focus. While the system deployment perspective specifies rough geometries and shapes of the overall system configuration, a geometry perspective could also be about specific geometries and shapes of subsystems. Current research is looking at systems from this perspective. The German FAS4M project [49] is one research project in that area, funded by the German government. Its goal is to bridge the methodological gap between functional architectures and a structural specification of the system that goes as far as to interlink system models in model-based systems engineering with geometrical and shape data as today used in the area of Computer Aided Design (CAD).

The *topology perspective* is one that is frequently encountered in the communication networks domain and several other areas. It looks at the structures of interconnections between system elements. As opposed to the system deployment perspective it will not focus on distances and precise locations. Instead of showing the distribution of museum robots and apps across the world, the system topology perspective would for example be concerned with the question how many different communication routes there are between the user's handheld application and the museum robot.

9.8   RELATION TO THE SYSTEM CONTEXT

9.8.1   Validity of the System Boundary

All perspectives mentioned in this chapter should be compliant with the system boundary that is defined by the system context (see Section 8.2). This means that the model elements belonging to the perspective should only represent aspects inside the system boundary. Interaction points for context interactions can exist in all perspectives, because they are "ports" that can have multiple aspects [88], which often cannot be addressed by one perspective alone. In general, the interaction points of one perspective are a subset of the interaction points of the complete system context, because not all aspects of all perspectives may be relevant for each of the interaction points. For example, it may not make sense to model the layer perspective for a mechanical joint to be connected to an external system. Section 9.9 will address the mapping between different representations of the same interaction point between different perspectives.

9.8.2   Using the System Context as part of the Stakeholder-Specific Views

Showing the system as a part of its context can help clarify the system architecture. An example is shown in Figure 9.17. The example shows

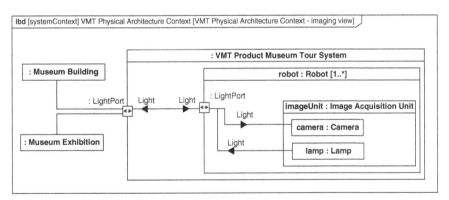

Figure 9.17. Example view on the Virtual Museum Tour system, taking parts of the system context into account.

how light propagates from the robot to its environment and is the used for the purpose of imaging. In this example, SysML internal block diagram is used to show the inside of a system context object.

9.8.3   Special System Context View for Verification

Verification people are stakeholders in system architecture. Their concern is the proper verification of the system. Even though verification should help ensuring that the system works in its utilization phase, the verification may use the system different from its typical utilization. For example, humans pressing a button may be replaced by machines that can cause well-defined stress conditions through high repetition rates of button presses. Human machine interfaces may be used by automata that can run different test case scenarios in a fraction of the time a human user would need for the same procedure.

The different utilization of the system in its test environment can lead to a system context that is different from the normal utilization context. This can be modeled by means of a different system context, which we call the *verification context*.

In the case of the Virtual Museum Tour System, one could think of a highly automated test environment, in which external influences like user commands to the handheld application or visual input to the robot's camera are emulated by automated equipment (Figure 9.18).[3] For the sake of measuring power consumption of certain subsystems, a lab power supply could be used instead of a museum's electrical installation—and all equipment could be controlled and monitored by means of a test control computer.

The special verification context according to Figure 9.18 can be modeled in SysML just like a normal system context (Section 8.2). The «systemContext» simply needs to be a different one, whereas the «system» block would be the same one as in the normal system context. This way a different context interaction of the system can modeled, as shown in Figure 9.19.

---

[3] Of course the initial cost for establishing such elaborate infrastructure only pays back with high market volumes or high unit prices. A company only making museum tour systems might not be able to afford it, but maybe a company that is world market leader in all kinds of robotic applications would already have such equipment in place and would also use it in cases it liked to develop and verify a museum tour system.

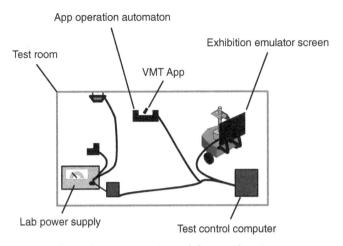

Figure 9.18. An informal representation of the verification context.

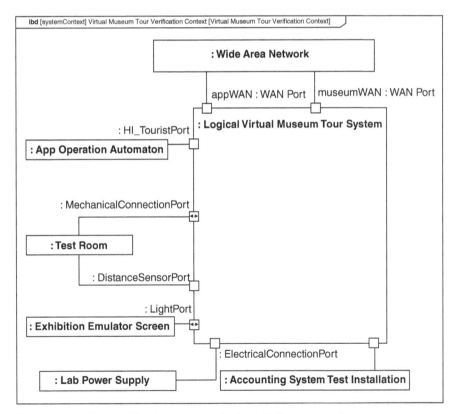

Figure 9.19. The verification context in SysML Representation.

9.9   MAPPING DIFFERENT PERSPECTIVES AND LEVELS

Mapping the perspectives to each other is an important aspect of a holistic understanding of the system. Remember that the real system only has one single architecture. The different perspectives enable the description of the system via different kinds of information, but in the end everything should line up in one consistent architecture to be exhibited by the system under development. This section describes how to map the different perspectives and levels in system architecture to each other in order to make the resulting architecture description an interconnected holistic description of the system as whole.

9.9.1   Functional-to-Physical Perspective Mapping

The mapping of functional blocks to physical blocks is called functional-to-physical mapping [53]. Ulrich [139] distinguishes different types of architecture, dependent on the kind of mapping: if each functional block is mapped to one physical block, he talks of *modular architecture*, whereas an approach of providing functions via a combination of different physical blocks is called *integral architecture* (Table 9.1). Different kinds of nonfunctional requirements may lead to different kinds of architecture. For example, integral architectures may be a suitable approach for microelectronics, because microchips can be made smaller if the same processing unit is used for contributing to different functions of the system of interest.

The mapping can be expressed by means of a table (allocation matrix) or it can be modeled. In SysML, functional-to-physical allocations are defined by means of «allocate» relationships that are drawn between the part properties in the functional architecture and those in the physical architecture.

Table 9.2 shows an allocation matrix of the Virtual Museum Tour System. Note that most SysML modeling tools can also create the table representation according to Table 9.2. A convenience feature of at least some of the SysML tools is to automatically create «allocate» relationships based on a double-click in the corresponding cell of an allocation

TABLE 9.1   Modular vs. Integral Architectures according to Ulrich [139]

Functional blocks to physical blocks relationship	1:1	1:N	N:1	N:N
Type of the architecture	Modular	Integral	Integral	Integral

TABLE 9.2 A Functional-to-Physical Allocation Matrix for Some Sample Elements of the Virtual Museum Tour System

	Tour management	Customer management	Exhibition management	Robot management
VMT cloud services	x	x	x	
VMT App	x			
Local VMT server	x	x	x	x
Indoor network access point	x			x
Robot	x			x
Robot charging station				x

matrix. Section 14.7 will show examples of a functional-to-physical mapping in SysML.

In practice, the pure allocation may not be enough to clarify the contribution of the different physical blocks to providing a function. This is particularly important to consider in the area of integral architectures. A more detailed system design may be needed to specify the way of mapping functions to physical blocks. One way of achieving this is to create relationships between functional blocks, physical blocks, and detailed requirements toward the physical block, specifying how the block has to contribute to providing a function. An example is shown in Figure 9.20. The figure also shows how to address nonfunctional requirements, which are at risk to be forgotten in a very functional-oriented approach. The nonfunctional requirements for various subsystems are derived from the system-level nonfunctional requirement. This is indicated in the model by means of the extended derive relationship «exDeriveReqt» that has been introduced in Section 7.1. In Figure 9.20, the detailed system architecture[4] is given by the part properties of the functional and physical perspective, the rationale element and the «allocate» and «exDeriveReqt» relationships.

---

[4] Sometimes we hear the term *system design* for the artifact we call "detailed system architecture" here. Since this book is based on the notion that system architecture and system design are synonyms, we are not making the distinction.

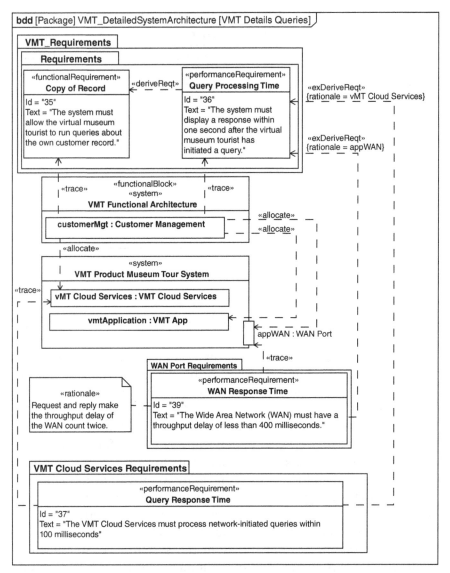

Figure 9.20. Detailed system architecture and traceability to requirements.

9.9.2   Mapping More Perspectives

As we have seen, the deployment perspective is based on system elements from another perspective. By using these system elements, a relation to the other perspective is automatically given.

From the behavioral perspective, we usually see the behavior of one or more system elements. Again, the relation to these elements is implicitly in their use within the views of the behavioral perspective.

The functional-to-physical mapping has already been described. What remains is to describe mappings in which functional, physical, and layered perspectives exist. The following has been based on Dänzer et al. [30]:

- «allocate» Relationships can be used to link the part properties that result from functional blocks with the ones resulting from «layer» blocks, where the direction should be the same as in functional-to-physical mapping, when looked at from the perspective of the functional block. It is recommended to split functional blocks in such way that the resulting blocks can each be linked to one unique layer. Functional blocks that have been created only for the corresponding reason of having a unique mapping to layers, and not based on domain knowledge, should be marked. To do so, we can assign the stereotype «layered» and concatenate the block name with the layer name. In the case of the "Robot Management" function running on the VMT server's application and technical service layer, we could for example decompose the functional block "Robot Management" into a block "Robot Management - Application Layer" and "Robot Management - Technical Service Layer."
- Finally, part properties that result from «layer» blocks can be linked with physical blocks by means of an «allocate» relationship, where the direction is the same as in the mapping to a functional block, seen from the perspective of the physical block.

9.9.3 Mapping Different Levels

In this book, the functional perspective and the logical and product level of physical architecture are described. In case all of them exist, a logical-to-product mapping can be made by means of SysML «allocate» relationships, as shown in Figure 9.21.

In certain cases, it may be beneficial to derive the product architecture from the logical architecture by means of a specialization relationship. In that case, the part properties of the logical architecture will be inherited into the product architecture. Then, those part properties to be defined more precisely in the product architecture can be redefined. This

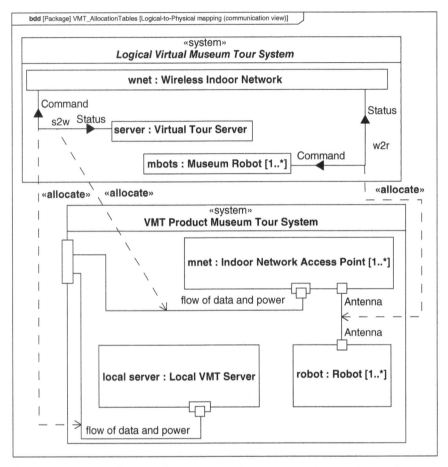

Figure 9.21. Logical-to-product mapping in a communication-focused view.

is particularly beneficial if parts of the logical architecture are so technically concrete that a 1:1 reuse into the product architecture is possible.

As always, there are some pros and cons to this approach:

- Pro specialization instead of «allocate»:
  - There is a tight connection between the logical architecture and the product architecture. The model ensures that the product architecture remains compliant with the cornerstones that are set by the logical architecture.
  - In case of overlap between the logical level and the product level, reuse of modeling is possible from one level to another.

  – The «allocate» relationship establishes a loose connection between the logical architecture and the product architecture, resulting in the potential for inconsistencies between both.
- Contra specialization instead of «allocate»:
  – The tight connection between logical architecture and product architecture makes changes in the logical architecture directly influence the product architecture.
  – Redefining elements of the logical architecture may occur so often that the overview is lost.
  – When using an «allocate» relationship, then the development of the logical architecture is decoupled from one of the product architecture.

9.10 TRACEABILITY

A special view is the traceability view. It ensures that the concern of finding the requirement behind an architectural solution can be addressed. The modeling of traceability has been discussed in Section 7.1.

9.11 PERSPECTIVES AND VIEWS IN MODEL-BASED SYSTEMS ARCHITECTING

9.11.1 Creating Different Views in a Model-Based Approach

A text document is sufficient to establish different perspectives and different views (document-based approach). For example, the functions in Figure 9.1 can be represented as a bullet list, preserving both the functional focus of the functional perspective and the hierarchical representation of the given view:

- System
  – Tour Management
    * Core Tour Management
    * I/O Tour Management
  – Customer Management
    * Core Customer Management
    * I/O Customer Management
  – Exhibition Management
  – Robot Management

If text is not sufficient, freestyle diagrams like Figure 9.1 can be used to complete the architecture description.

The model-based approach is thus not necessary in order to produce sufficient architecture documentation. Conversely, this means that model-based systems architecting will have to compete with intuitive-to-read documentation that can be created by means of freestyle writing and drawing. The system architect should thus strive for generating views that are as close to the stakeholders' preferred representation as possible. In the case of Figure 9.1, it is obvious that the native SysML representation according to Figure 9.2 will most likely satisfy the stakeholders equally well. A good modeling tool can create figures like Figure 9.2 automatically, once the displayed information has been entered in the model. Some tools even offer automatic layout of diagrams.

In case the stakeholders' preferred representation cannot be produced within the possibilities of the modeling language or the used tool, reporting and document generation can be considered. The more specialized the views need to be the more difficult or costly it will be to generate them from a model. In order to justify the resulting cost of model-based systems architecting, one should thus be aware of its benefits.

On the benefit side, the model-based approach offers means of ensuring consistency between different perspectives and views in an effective manner. For example, if "Customer Management" needed to be renamed to "User Management" in the earlier example, a model-based approach would only require one renaming action in the model, and all different views would automatically use the term "User Management" from the moment of the first renaming on. Model-based Systems Architecting thus increases efficiency and consistency in documenting system architecture with different perspectives that serve different purposes and different architecture views that address different stakeholders' concerns.

In balancing cost against benefits, the system architect should thus find the appropriate trade-off between serving the stakeholders with their preferred representation of information and keeping the needed infrastructure (modeling tools, reporting tools, etc.) affordable and maintainable. In case of doubt, we recommend to keep it simple, this means to try avoiding complex infrastructure. In our experience, stakeholders are more reluctant to work with information from a complex IT system than to accept a representation of data that is not 100% according to their preference. This may be due to the fact that complex IT infrastructure has less availability[5] than simple tools, and availability of data is an important success factor in business.

We often heard stakeholders, but also architects justify freestyle architecture documentation by pretending that the consistency can be easily maintained without a modeling tool. However, we also saw how much confusion can be created if inconsistent names are being used for elements in the system—or even worse: if the same term is used for two different things, for example by naming a subassembly exactly like one of its parts. When such confusion is discovered, it has often spread out across the stakeholders, in large organizations even throughout different countries and documentation repositories. The cost of correcting such problematic use of terms can be moderate to high, if just the effort for the correction of documents is measured. However, the real damage caused by inconsistent terminology can only be assessed if more soft factors like conflicts arising from misunderstandings and stakeholders' effort for learning corrected terminology are taken into account.

We therefore recommend the model-based approach together with the notion that the model is the single source of truth. Team exercises like the joint creation of a domain knowledge model according to Section 8.5 in a workshop can help the stakeholders understand how ambiguous their use of terms is if not consolidated in one model.

9.11.2 Using SysML for Working with Different Perspectives and Views

The different diagrams in the previous sections show examples of representing different views in both informal notations and in SysML. Some concrete hints for organizing models with different perspectives in SysML are given in Section 7.9. Here, we discuss how SysML can address the different perspectives and how to provide suitable views for different stakeholders.

It can be seen that SysML can represent tree-like breakdown structures very well (e.g., in Figure 9.2). It can also represent allocation matrices where the mapping of different perspectives is described.

What engineers would often describe in their jargon as "block diagrams" can be provided by the SysML internal block diagrams (Figures 9.4 and 9.6). Interfaces can be shown on a high abstraction

---

[5] If one tool has a probability of unavailability of p, then a tool chain of N tools of equal availability have the probability of $N \cdot p$ under some typical assumptions.

level by means of internal block diagrams and then refined, either by specifying details of data flows via a domain knowledge model (Figure 9.5) or by adding details about ports to the specification of their port types (Figure 9.11). We highly encourage system architects to use these SysML notations for interface definitions during workshops with the corresponding stakeholders, because a system architect should have a commonly understood and agreed specification of interfaces as one key objective (Section 11.2.1). SysML helps creating well-understood interface specifications by being both precise and intuitive.

In case the stakeholders are not familiar with SysML, particularly the approach via a domain knowledge model is an intuitive one that can be used after a short explanation of the involved SysML syntax. Only few SysML users are so skilled with a modeling tool that they can use it live in a workshop with stakeholders. If a version of such a diagram already

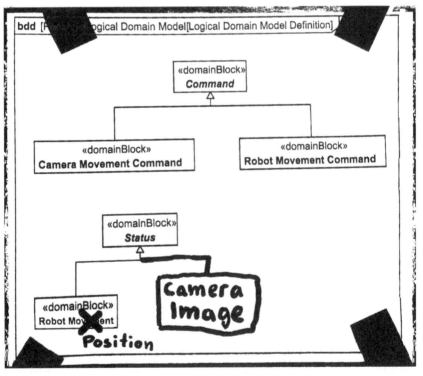

Figure 9.22. The partially hand-made version of Figure 9.5, as it could have come out of a workshop with stakeholders of the museum robot's higher communication layers.

exists prior to the workshop, then we usually print it on a poster and bring it to the workshop, allowing the stakeholders to draw on it or put sticky notes onto it. Figure 9.22 shows how Figure 9.5 would have looked like if it had been created during such a workshop.

In many cases, we would recommend using flipcharts and markers for creating domain knowledge models like the one in Figure 9.5. For all those who think that flipcharts are suitable modeling tools for parts of their daily modeling use cases, we recommend the book "Agile Modeling" by Ambler [8].

Chapter 10

Typical Architecture Stakeholders

10.1 OVERVIEW

Stakeholders in the context of system architecture are the main focus of this chapter. However, we first need to set a larger context, because stakeholders are a topic in requirements engineering as well.

Requirements toward the system of interest lead to concerns in system architecture. Requirements engineers capture requirements in direct dialog with people or entities they call "stakeholders" or "stakeholder representatives" in a requirements engineering context. The users of the system are very important stakeholders in requirements engineering, and also acquirers, regulatory bodies and entities performing marketing, production, training, operations, and maintenance will typically have requirements toward the system of interest. It is a matter of taste whether these people or entities are explicitly identified as stakeholders in a system architecture context or whether the system architect will consider the requirements engineering people to be the actual architecture stakeholders when the concern is to satisfy requirements.

Apart from their role as potential providers of requirements, some people or entities may have more interaction with the system architect and more concerns to be addressed during systems architecting than

Model-Based System Architecture, First Edition.
Tim Weilkiens, Jesko G. Lamm, Stephan Roth, and Markus Walker.
© 2016 John Wiley & Sons, Inc. Published 2016 by John Wiley & Sons, Inc.

covered within the scope of the requirements engineering processes. They are the ones who should collaborate with the system architect, which makes them potentially important stakeholders in the context of system architecture, no matter whether they have requirements toward the system of interest. This also includes people who are no stakeholders in the context of requirements engineering but have a stake in the system architecture.

The further sections in this chapter are about this latter kind of stakeholders. Selected aspects of the system architect's collaboration with them will be discussed. Each of the following sections is dedicated to one typical architecture stakeholder inside a typical organization and the organizational interface to the system architect. Depending on the nature of the business and the involved organizations, the described entities may exist as distinct departments or as persons with certain roles or tasks. They may exist in the organization in which the system architect works or in one of its business partners. The selected stakeholders and topics are not generally applicable, because they are examples. You may consider using this chapter as an inspiration when analyzing who are the stakeholders of your own system of interest.

We describe the collaboration between the system architects and the selected stakeholders as a win-win situation, because the stakeholders may be looking at their own wins when asked to reserve time for cooperation with the system architect. The tables at the end of each of the following sections characterize the cooperation between the system architect and the respective stakeholder in showing what each of the parties has to give and what can be gained in return.

This chapter covers stakeholders that are often encountered for typical systems of interest. It should be noted though that such a system is typically surrounded by *enabling systems* (refer to ISO 24748-1:2010 [61]), which are used, for example, during development, manufacturing, maintenance, and retirement of the system. The enabling systems have to be compatible with the system of interest and have interfaces with it, that is, the external interfaces of the system. These interfaces will typically be found during the work on the system context (Section 8.2). The people concerned with the enabling systems may be additional architecture stakeholders for the system of interest. They are not comprehensively discussed in this chapter, because who they are and how they interact with the system architect is very much dependent on the system of interest and the organizations involved in its life cycle.

10.2   REQUIREMENTS ENGINEERING

The role of requirements engineering as a stakeholder of systems architecting has been briefly discussed in the previous section. Requirements engineering people must communicate the requirements to be satisfied by a certain system architecture to the responsible system architecture people. "Communicate" means ensuring that mutual understanding is created by means of at least partly verbal direct and bidirectional communication.

Systems architecting results in system architecture that satisfies the requirements. To check if the requirements have been correctly understood by the individuals involved in systems architecting, the system architect should communicate with the requirements engineer. In this case, the communication ensures that an input to system architecture has correctly been received.

If requirements on a lower abstraction level need to be derived from the system architecture, there is also an output from the systems architecting discipline towards the requirements engineering discipline. There is thus a bidirectional communication link between requirements engineering and systems architecting roles in cases with multiple abstraction levels. It directly corresponds to the relationships between requirements and architectures in the SYSMOD zigzag pattern we saw in Section 7.1. Figure 10.1 illustrates the bidirectional nature of the interface between requirements engineering and systems architecting: The upper handshake in the picture means that requirements engineering

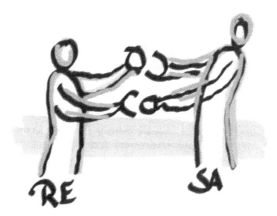

Figure 10.1. Collaboration between requirements engineering (RE) and systems architecting (SA).

TABLE 10.1 Close Collaboration between System Architects and Requirements People as a Win–Win Situation

What requirements engineering people give	• Explanation of the requirements
What they get in return	• The requirements are being taken in for further processing • The system architecture is explained to them for deriving requirements on a lower abstraction level
Obligations toward system architects	• Requirements people hand over requirements to system architects and ensure that a common understanding is reached.
What they can expect from system architects	• System architects account for the requirements in systems architecting. • System architects give feedback if requirements are not clear or deficient. • System architects ensure that system architecture descriptions are traceable to requirements. • System architects provide architecture descriptions that can be used for deriving requirements on a lower abstraction level.

provides requirements as an input to systems architecting. The lower handshake means that systems architecting provides an architecture description as the basis for deriving requirements on a lower abstraction level. The picture has been provided by Tim Weilkiens [146], who made it for blogging about the need for close cooperation between requirements and architecture people and about the two directions of communications that are implied by the SYSMOD zigzag pattern .

Table 10.1 describes the cooperation between system architects and requirements people as a win–win situation.

10.3 VERIFICATION

Verification is supposed to show that the system has been created according to its specification. According to the V-Model[1], for example, in a version according to Emes et al. [36], *system tests* verify the fulfillment of system requirements, and *integration tests* verify the fulfillment of architectural specifications. Tests can be replaced by other verification approaches like inspection or demonstration. In any case, the expected nature of the system is derived from its specification, and compared with the observed nature from verification. This activity requires several inputs that are either work results of systems architecting or can be derived from these work results:

- For assessing the fulfillment of architectural specifications the people responsible for verification need the system architecture description. The system architect needs to communicate with them in order to ensure common understanding.
- The system architect can support the appropriate design or assignment of priorities in verification. For example, if the risk of failure in a certain state of the system of interest needs to be minimized, the system architect can provide views showing how the system can enter the given state. On this basis, test cases provoking the system to enter the given state can be designed or prioritized.
- Methods based on the so-called equivalence classes aim at coverage of system requirements with tests. They avoid the typically unpractical or even impossible attempt to cover the full input vector space of the system with test vectors. Instead, they recommend the use of test vectors that cover a set of "equivalence classes" of the input vector space with one test vector per equivalence class. The work by Richardson and Clarke [120] and the category partition method [111] in software testing have enabled system-level methods like classification trees [24]. Richardson and Clarke [120] propose looking at both the requirements and the solution when designing tests for coverage. This indicates that the design of system tests may require knowledge about the solution chosen for the realization of the system of interest, that is, its system architecture. Practically this means that the people who plan or design system

---

[1] A description of the V-Model can be found in the Appendix B.

tests are stakeholders of system architecture and that the system architect should closely cooperate with them.

- After changes in the system, regressive verification needs to assess that the changes produced the desired effect and did not produce side effects in the system. Only by having a thorough understanding of the system of interest and its inner connections and dependencies can one assess which steps need to be repeated during regressive verification. In other words, the knowledge of the system architecture is required in the design and planning of regressive verification. An easy way to make the required architecture knowledge available during the verification planning process is to involve system architects with knowledge about the appropriate parts of the system architecture in that process.

Another aspect of system architecture involvement in verification activities is the postprocessing of failed verification. Particularly in system testing, the cause for a failed test step can potentially reside in multiple parts of the system (if not in the test procedure itself). Based on the system architect's expertise about the system structure it is easier to narrow down the number of hypotheses about the potential cause of a failure.

People with responsibility in verification are thus stakeholders of the system architecture and should have a strong interface with system architects.

Furthermore, the system has to be architected for verification. This may mean to make the system easy to inspect for verification by inspection. This also means that systems architecting should account for the fact that a system needs to be tested during its life cycle. For example, testability requirements may lead to system functions like dataloggers or to additional interfaces like test access points. In Section 9.8.3, we have seen that it is possible to show a variant of the system context diagram that describes the system in its test context. Test access points should be part of such a context diagram, and the diagram should explain what is connected to them during the verification of the system. The detailed modeling of the connections with test access points allows for a precise interface specification.

Table 10.2 describes the cooperation between system architects and verification people as a win–win situation.

TABLE 10.2   Close Collaboration between System Architects and Verification People as a Win–Win Situation

What verification people give	• Trust in the system architect's expertise about the system • Expertise on the intended ways of verifying the system
What they get in return	• The system is designed for verification. • Verification design and verification planning is based on information about the solution. • Regression testing is based on the assessment of impacts of changes.
Obligations toward system architects	• Verification people involve system architects in verification planning. • Verification people involve system architects in the analysis of failed test steps.
What they can expect from system architects	• System architects provide expertise about the system of interest.

10.4   CONFIGURATION MANAGEMENT

This section has nothing or little to do with the configuration of variants according to Chapter 15. Configuration management in the context of this section ensures that versions are tracked across different deliverables in system realization. These deliverables are called *configuration items*. They can be documents, models, and implemented or assembled elements of the system. Configuration management defines *baselines*, that means sets of versions that are considered the state of the configuration items at a certain moment or decision gate. It also tracks the compatibility between different versions of different configuration items [80].

The system architect can help identifying the different configuration items in the system, based on the system models, for example, of the physical perspective. Models in model-based systems architecting can

TABLE 10.3   Close Collaboration between System Architects and Configuration Managers as a Win–Win Situation

What configuration managers give	• Explanation of the applicable configuration management strategy • Overview of system configurations
What they get in return	• Versioned system architecture description deliverables into baselines • An overview of the configuration items in the system and their compatibility
Obligations toward system architects	• Configuration managers explain the needs of configuration management. • Configuration managers plan baselines together with system architects.
What they can expect from system architects	• System architects deliver system architect's configuration items into baselines on time. • System architects offer consulting services regarding the definition of configuration items and the assessment of compatibility.

be used as the single source of truth for the name of the different configuration items. The system architecture description can facilitate the assessment of compatibility between different versions of elements of the system, based on, for example, interface specifications.

The system architecture description is a configuration item as well and needs to be versioned. In model-based systems architecting, one needs to decide whether the model itself or a set of views generated from it is entering a configuration baseline. In any case, views should be traceable to the version of the model from which they were created.

Table 10.3 describes the cooperation between system architects and configuration managers as a win–win situation.

10.5   ENGINEERING DISCIPLINES

The engineering disciplines produce work results that will in the end make the system exhibit the desired architecture. It is therefore necessary that the engineering disciplines are committed to the system architecture in order to make it materialize. An architecture description has value only if it has been created with stakeholders from the engineering disciplines that realize the system's subsystems.

In a close dialog between system architects and engineering disciplines, a common understanding about the system and its principles of operation is obtained. Ideally a multidisciplinary dialog is established and maintained throughout the whole specification and implementation cycle. The system architect takes the lead in maintaining the dialog, mediating during trade-offs and securing that work results are documented, validated, and made reproducible.

Typical representatives of engineering disciplines are:

- Division heads or group leaders
- Subsystem architects (e.g., software architects in the case of a software engineering disciplines)
- Developers

The more hands-on knowledge the involved development stakeholders have, the more likely the resulting system architecture can be followed smoothly during implementation. The system architect should have an opportunity to talk to the developers, who are allocated to the actual implementation work and not only to their leaders.

A particularly important aspect of the architect's work is to ensure that interface agreements are made, documented and followed. This includes both interfaces between different subsystems as well as flows of information, materials, force or energy between different functional elements within the same subsystem.

The negotiation of an interface agreement has to be made with the stakeholders on both sides of the interface. For example, an interface between the battery charger of the museum robot and the robot itself has to be negotiated with both the stakeholder representatives responsible for the battery charger and stakeholder representatives responsible for the robot.

It is important that engineering disciplines acknowledge the need to follow interface agreements or to inform the system architect if adhering to the interface turns out to be impossible or discouraged. This seems to be trivial, but our practical experience has shown that stakeholder representatives who are for the first time collaborating with a system architect sometimes consider their job done when the first version of an interface agreement has been made. They might in consequence ignore the change processes that need to be followed when changing an interface definition. We have observed two patterns of violating interface agreements:

- People working on the element on one side of the interface change the interface definition without notifying the people with the responsibility for the element on the other side of the interface, which leads to improper system performance on integrating the elements on both sides of the interface during system integration.
- People working on the elements on both sides of the interface change the interface definition together, but without involving the system architect, which leads to a discrepancy between the system architecture description and the architecture that is exhibited by the system and in consequence often to side effects that the people who changed the interface definition were unaware of.

We therefore recommend to system architects to periodically check if interface agreements are still followed within the development organization, particularly if system architecture is new to the involved stakeholders.

It is important that system architects and stakeholder representatives from the engineering disciplines establish a mutual trust relationship. A challenge in that regard is the fact that the system architect has to interfere with the solution in the different disciplines in order to optimize the overall system performance. Stakeholder representatives from the engineering disciplines may consider this an act of interfering with something that is not the system architect's business. It is necessary for the engineering domains to understand the need for the system architect to influence how the work in the disciplines is being done - but it is also necessary for the system architect to know where the own competence ends. Particularly system architects who have been recruited from within the organization often have a background from one of the involved fields of engineering and have a particular difficulty not to get involved into the corresponding discipline's own business.

A successful introduction of systems architecting in an organization means that the mutual trust between system architects and engineering domains is established and both sides understand that the boundary between system architecture work and domain engineering is unsharp and has to be sharpened on a case-by-case basis in the mutual dialog. It is a measure for the success of introducing systems architecting that engineering disciplines and system architects do not blame each other for exceeding their competence, but solve boundary questions together without any need for escalation.

We see three different scenarios how system architects and people from the engineering disciplines can collaborate:

- The system architect leads the definition of the system architecture.
- The system architect has to run after architecture decisions that have already been made explicitly or implicitly by the concepts the engineering disciplines have based their work on. The system architect has the license to document them and acts as system archeologist.
- The system architect and the stakeholders from the engineering disciplines see themselves as a team that acts in a self-organizing way. Usually, multiple such teams should be set up around different aspects of development, for example, around a function or a set of functions of the system. Sometimes, such teams are known as feature teams, but then their scope usually comprises more than just the definition of the system architecture.

The above scenarios are not mutually exclusive, that is, often the real collaboration is a mixture between them.

Table 10.4 describes the cooperation between system architects and people in different engineering disciplines as a win–win situation.

10.6   PROJECT AND PRODUCT MANAGEMENT

System architects can use their knowledge about the system of interest to help the project manager in achieving realistic plans and thorough risk management. The system architect's knowledge about the system structure, interdependencies inside the system (Figure 10.2), and the complexity in different areas of the system are the key enablers in that regard.

Here is a list of typical tasks in a project to which a system architect should contribute in close cooperation with project management or the people to whom the tasks are delegated by project management:

- development and verification planning
- system integration planning
- technical feasibility analysis
- work breakdown
- analysis of technical dependencies between development activities (Figure 10.2)
- risk management
- effort estimation.

TABLE 10.4 Close Collaboration between System Architects and Engineers in the Disciplines as a Win–Win Situation

What engineers in the disciplines give	• Expertise about their domain • Willingness to change the design of a subsystem to make the whole better • Trust that the system architect will respect their own domain's competence.
What they get in return	• Their stakeholder view toward interfaces with other subsystems will be respected by other stakeholders, for example, engineers working on another subsystem. • Their needs will be taken into account. • They will be trusted that they are competent within their own field.
Obligations toward system architects	• Domain engineers report back if interface agreements cannot be met. • Domain engineers have system architects lead the change of interface definitions instead of changing interfaces alone.
What they can expect from system architects	• System architects keep domain engineers updated about system architecture changes and involve them in system architecting and change activities that affect them.

The risk management activity is one that is closely related to the system architect's work. Trade-offs in system architecture often trade risk against effort. For example, the decision how generic an interface should be trades the high effort for making a generic interface against the risk of late changes in case an interfaces is initially designed for very few use cases only and needs to be extended upon discovery of a forgotten or newly required use case. The system architect does not own the budget that tells how much effort can be spent and therefore needs to enter dialog with the stakeholder who does own it. The project manager is often the stakeholder to involve in that regard. Our advice to the system architect is to not only report risks, but also proposals for mitigation strategies and the system architect's own preference among those. A more general discussion about risk management is due later, in Section 17.3.

Figure 10.2. Technical dependencies © 2014 Jakob K., reproduced with permission.

In the context of effort estimation, the system architect also has to estimate the effort for the tasks that will need to be done within system architecture.

Of course, the work-load for system architects depends on the nature of the project, but also on the concrete role description of the system architect in the given organization.

If the role description stays the same across multiple projects and across time, it is worth making a book keeping about the effort spent on different projects. This will improve estimates over time.

An example is shown in Figure 10.3: it shows a fictitious curve of the system architect's workload over time for systems architecting in a fictitious project and for collaboration with the stakeholders, as discussed throughout this chapter.

The curve in Figure 10.3 has two peaks:

- The first peak could relate to the system architect's activities that are involved in designing the system.
- The second peak could relate to the system architect's activities in connection with verification (Section 10.3), but also to topics that occur during preparation of production and to change management activities that result from learnings after integrating close-to-final products toward the release.

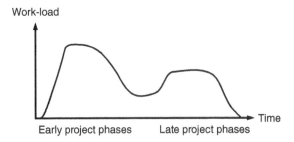

Figure 10.3. A fictitious example of predicting architecture work-load over time.

Even though the curve in Figure 10.3 is fictitious, similar curves have been observed in real industrial projects.

Project leaders and system architects are two poles of a project: while project leaders have to drive the project forward and keep timelines, system architects have to ensure that technical work has enough technical depth and maintainability to ensure the required quality over the whole life cycle of the system architecture, not of the project. The project managers "pull forward," the architect "pulls down" (Figure 10.4).

One could deduce that a project driven by a project manager in close collaboration with a system architect would reach sufficient technical

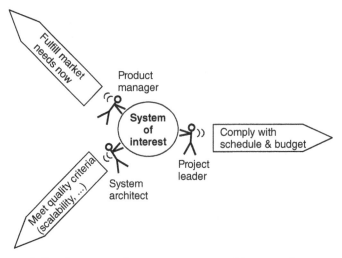

Figure 10.4. Split of concerns between system architect, project manager and product manager.

perfection within the right time, but there would be a risk that it would produce technology for its own sake ("happy engineering"). To ensure that market and customer needs are accounted for, someone should "pull up," that is, ensure that the market needs are taken into account. This is a task of stakeholders like requirements engineering or product management. Typically, requirements management has captured post-processed stakeholder input, and the architect will get the stakeholder view from the requirements, or – better – from a requirements engineer who ensures that the requirements are understood and accounted for. Yet it may be necessary to bring the project leader, the system architect and product management closely together in order to run fast iterations between pulling up, down, and forward according to Figure 10.4 and thus bring a project on the right track. To keep the project moving into the right direction, they should stay in close contact during the whole duration of the project.

Table 10.5 describes the cooperation between system architects and project leaders as a win–win situation.

10.7 DEVELOPMENT ROADMAP PLANNERS

Development projects have a defined start point in time, an expected end point in time, and an expected deliverable. The deliverable can for example be a concept, a product, a subsystem or a new feature. Usually, several development projects can run in parallel. For larger organizations, there is typically such parallelism at any moment, whereas very small organizations may be in situations in which one project loads their full capacity. Even in the latter case, parallelism may occur during the transition phase in which one project ends and another one starts.

There should be a list of both current and future development projects that provides overview about their assumed start dates, end dates, and deliverables. We call it the *development roadmap*. For simplicity, let people who are responsible for maintaining this list be called *roadmap planners*. In different organizations, this can be single persons or organizational entities.

System architects and roadmap planners have several good reasons to cooperate. Of course roadmap planners should also collaborate with other stakeholders in order to complete their puzzle. Here we describe the puzzle piece that should be addressed between the roadmap

TABLE 10.5 Close Collaboration between System Architects and Project Leaders as a Win–Win Situation

What project leaders give	• Time of their engineering resources for maintaining the multidisciplinary dialog between engineering and system architects • Trust in the return of investment that will be generated when investing into good system architecting
What they get in return	• Overview and clarity about the project's deliveries • Improved communication in the project • Analysis of dependencies (Figure 10.2) • Predictability of cost and performance • Early information about risks • Better products (see Chapter 3)
Obligations toward system architects	• Involve system architects in planning in order to account for the correct dependencies (Figure 10.2) • Involve system architects in decisions that affect the system structure, system functions, or system behavior • Ensure that the time for proper systems architecting is allocated on the system architects schedule and on the schedule of the architecture stakeholders.
What they can expect from system architects	• System architects offer consulting services during work breakdown, planning, technical feasibility analysis, and risk management. • System architects provide overview documents and diagrams that help scoping the project. • System architects immediately report if they perceive that plans are unrealistic or that the risk profile needs to be updated.

planners and the system architect. In our eyes, the most important aspects of this collaboration are as follows:

- System architects have an overview of existing system architecture and can thus extrapolate their knowledge to new projects. This enables rough effort estimates and a statement whether the development roadmap looks realistic in terms of technical feasibility and time to completion.
- System architects can analyze high-level dependencies between projects like: "Subsystem X only runs if subsystem Y is installed. We better complete the first assemblies of subsystem Y well before the first scheduled prototype test of subsystem X, or alternatively have to build an emulator of subsystem Y."
- System architects need to plan their own work, which is usually feeding into the different development projects on the development roadmap.

Let us elaborate on the last aspects of work planning for system architects. Each project means a certain work-load for system architects. We saw in Section 10.6 that one can make a prediction of the work-load for a given project (Figure 10.3). The roadmap enables the creation of an overview of work-load over time, which may allow for optimizing resource allocation and planning in each of the development projects, with the aim of not exceeding the total capacity of system architects. This is particularly relevant in organizations in which system architects come from one team or department with a defined head count.

Figure 10.5 shows an example of such a work-load overview across projects. Some of the shown work-load curves are similar to the one

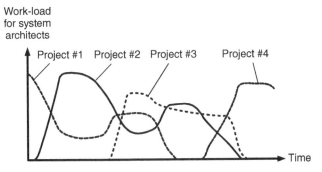

Figure 10.5. Using workload predictions from Figure 10.3 to support planning based on roadmaps.

from Figure 10.3 in Section 10.6. They are however not completely identical to that curve. The reader can easily verify that the total work-load for system architects would have high peaks at certain moments, in case the timing of projects or the shape of the different work-load curves in Figure 10.5 changed.

If we assume that the average work-load of project #1 to #4 in Figure 10.5 is perfectly matched to the capacity of available system architects, then the figure shows a close to ideal situation, in which peaks in total work-load are moderate because peaks and valleys of the individual projects' work-load curves level each other out. For example, the first peak load in project #2 coincides with a valley in project #1.

Reaching a close to ideal situation as described hardly happens by itself. Only by planning based on roadmap information can the system architects avoid capacity overflow or underflow. The easiest means of optimizing plans is to start tasks earlier than necessary or to postpone them if feasible with an acceptable risk. However, there may be cases in which it is impossible to resolve bottlenecks with planning only. The most easy way out in theory would be to add resources, but this is usually not what can be done in practice. Then two possible actions to consider are as follows:

- Suppress certain tasks or reduce their ambition level. This results in a risk that should be addressed via risk management as it has been discussed in Section 10.6.
- Escalate the bottleneck to the roadmap planners. It may be that other disciplines have bottlenecks as well and that roadmap planners have to admit that their roadmap is unrealistic at the end. This will then ideally result in a modified roadmap.

Table 10.6 describes the cooperation between system architects and people in roadmap planning as a win–win situation.

10.8 PRODUCTION AND DISTRIBUTION

Yassine and Wissmann [150] point out the strong interrelation between product architecture, assembly and distribution: the product architecture enforces certain assembly steps and enables or inhibits the delay of assembly to entities inside the distribution chain. This is why production and distribution people are important stakeholders in system architecture.

TABLE 10.6   Close Collaboration between System Architects and Roadmap Planners as a Win–Win Situation

What roadmap planners give	• Transparency about the roadmap creation and about roadmap changes • Willingness to take system architecture knowledge into account.
What they get in return	• More realistic roadmaps
Obligations toward system architects	• Roadmap planners provide early drafts of roadmaps and inform about later roadmap updates.
What they can expect from system architects	• System architects make rough feasibility studies based on roadmap input. • System architects optimize plans in system architecture based on roadmaps. • System architects escalate capacity bottlenecks that cannot be overcome by optimizing system architecture plans alone.

There are cases in which the production and distribution environment is not defined completely during product development but is already present to a certain extent when development starts. In such a case, it may be a constraint on the system architecture that certain production or distribution steps have to be carried out using given elements of an existing production and distribution environment. Such constraints are ideally captured by the requirements engineer and can be refined in the dialog between the system architect and the production and distribution people.

Table 10.7 describes the cooperation between system architects and people in production/distribution as a win–win situation.

10.9   SUPPLIERS

Suppliers provide deliverables that usually have interfaces with the system of interest or even inside it. The work with suppliers thus always involve interface specifications. Since system architects are the owners of interface specifications, they should be involved in the interface

TABLE 10.7   Close Collaboration between System Architects and Production and Distribution People Described as a Win–Win Situation

What production and distribution people give	• Insights into possible assembly scenarios • Insights into the assembly possibilities inside the distribution chain
What they get in return	• Products that are designed for manufacturing • Products that are designed for optimized distribution logistics
Obligations toward system architects	• Production and distribution people explain the processes they are responsible for. • Production and distribution people evaluate cost of different manufacturing and distribution scenarios.
What they can expect from system architects	• System architects account for manufacturability from the first moment of systems architecting • System architects take special needs of distribution or decentralized finishing of the manufacturing process into account

agreements with suppliers, who best also make their respective architects part of the dialog.

Table 10.8 describes the cooperation between system architects and supplier's architects as a win–win situation.

10.10   MARKETING AND BRAND MANAGEMENT

Yassine and Wissmann [150] have made a thorough analysis of the relationships between product architecture and marketing as well as brand management, among others. They come to the conclusion that product architecture influences the firm, thus the organizational entities that we would call stakeholders. In addition, they point out multiple open research questions existing toward the "consumer perspective regions." We can thus on the one hand consider marketing and brand management as potentially important architecture stakeholders but on the other hand not simply answering the mentioned open research

TABLE 10.8 Close Collaboration between System Architects and a Supplier's Architects as a Win–Win situation

What the supplier's architects give	• Technical insides from the supplier side that are needed for making well-founded interface agreements
What they get in return	• Technical insides from the customer side that are needed for making well-founded interface agreements
Obligations toward system architects	• The supplier's architects strive for a technically sound interface agreement
What they can expect from system architects	• The system architects strive for a technically sound interface agreement

questions. Therefore, we will limit ourselves to providing hypotheses based on our own experience.

Quality criteria are an important input to systems architecting, as it will be discussed in Section 12.1. We believe that the marketing and branding strategy of the organization provides important quality criteria, because it can answer how important criteria like easy product differentiation, short time-to-market, and the possibility of radical innovation are considered to be for sustainable success and growth on the market. The derived quality criteria may be important for the long-term success of a chosen architecture, whereas requirements-based making of architecture decisions only focuses on currently identified requirements. Since marketing people are usually focused on current market needs, the system architect should explain well what the life time of the system architecture is. It may need to take future market development into account. The architect should also be aware that nobody can predict the future. By making statements about the future, marketing will do guesswork and may change their mind whenever future market needs require this. The architect should always be prepared for change. The dialog with marketing people helps them to develop a sense for the needed flexibility over time, and the architect should use his own sense of the matter to make architecture decisions that reach further out into the future than marketing will be able to reach with their predictions.

The architect should also be aware of the very different fields. Marketing and branding people will not speak the same language as system architects. The system architect needs to explain the own needs well and should ask the right questions (e.g., "Is it more important to achieve strong differentiation of products or fast time to market?"). The system architect should also be aware that particularly marketing uses a language that may seem strange to people with a strong focus on the solution domain. This has multiple reasons, one being that marketing has to influence potential buyer's emotions, whereas the human factors usually considered in systems architecting are the ones important for human machine interaction.[2] As goes with each dialog, the one with marketing and branding people will only succeed if there is recognition for each other's work. In order to initiate a successful dialog, the system architect needs to accept that a more emotional and less technical approach to the system of interest is necessary for business success and is chosen by the marketing people on purpose and not due to lack of technical insides.

Table 10.9 elaborates on the above hypothesis by describing a potential win–win cooperation situation.

10.11 MANAGEMENT

Fried and Hansson write in their book "Rework" [42]: "if you're opening a hot dog stand, you could worry about the condiments, the cart, the name, the decoration. But the first thing you should worry about is the hot dog" (p. 72). Applied to systems architecting, this could mean: the first thing system architects should worry about is system architecture. Of course a system architect who has never facilitated an interface agreement and never produced any architecture documentation is like a hot dog seller without hot dogs. Still system architects need to communicate their work results not only to the development organization for ensuring that the realized systems exhibit the intended architecture, but also to their management stakeholder in order to ensure that the value of systems architecting keeps being recognized on management level.

Since most organizations do not sell architecture, the architects will need to justify how they contribute to the turnover of the organization.

---

[2] It would be a very nice discussion to assess whether the user's emotions need more focus in human factors engineering. This question is beyond the scope of this book, but we direct the interested reader to the article "Engineering Joy" [51], which we have already made reference to in Chapter 3.

TABLE 10.9   Close Collaboration between System Architects and Marketing or Branding People Can Be a Win–Win Situation

What marketing or branding people give	• Willingness to face a more technically oriented conversation partner than usual
What they get in return	• A system architecture that will support their strategy instead of just the next product on the roadmap
Obligations toward system architects	• Marketing and branding people explain the marketing and branding strategy. • Marketing and branding people answer the system architects' questions that help the identification of quality criteria for the system architecture.
What they can expect from system architects	• System architects accept that marketing is needed for business success, even if it speaks a language that sounds strange to some engineers. • System architects take the marketing and branding strategy into account in the daily work on system architecture.

Chapter 3 explains this and also gives hints on which values to communicate. It is important to communicate the value of systems architecting and the reasons to believe in return of investment to management on a regular basis in order to maintain support for architecting activities in the organization.

Depending on how "highlevel" the management likes to be addressed it can be sufficient to communicate and explain the value as described in Chapter 3 – best accompanied with concrete success stories from daily business that underline how system architects created value and return of investment for the organization. For the more technologically interested management stakeholders, it may however be as well important to show architecture deliverables like a piece of architecture documentation, again best associated with a success story indicating, for example, how this concrete piece of documentation helped systematically analyzing possibilities to reduce cost. Model-based systems architecting can assist the latter kind of communication, because it supports the creation of stakeholder specific views that can present model contents in a representation suitable for management.

It is our experience that effective systems architecting requires management awareness, even better: management commitment. On the way to get there, we have observed in organizations without well-established architecture practices that unofficial architecture activities by few engineers can produce first success stories of applying methods in system architecting, which will then serve as show cases for motivating the official introduction of systems architecting via a change project. In such early activities, it is particularly important to focus on "quick wins," that is, activities with low effort that can create high value that is visible to the organization.

Since management trust in the system architects is very important, system architecture people with the possibility or even the obligation to occasionally present status to management should reserve enough time for preparing and doing the communication to management.

TABLE 10.10   A Good Relationship between System Architects and Management as a Win–Win Situation

What management gives	• Trust in the return of investment that will be generated when investing into good system architecting
What they get in return	• Predictable projects • Better products (see Chapter 3) • Better communication between engineering domains
Obligations toward system architects	• Management is patient regarding return of investment.
What they can expect from system architects	• System architects explain the value of systems architecting (see Chapter 3). • System architects provide "bird perspective" information about the system of interest instead of overwhelming management with technical details.

Our three golden rules to consider if you are an architect who is about to meet management:

- When presenting problems always come with a solution proposal – or even better: an evaluation of several solution scenarios and a recommendation which one to choose.
- Use simple communication and do not expect management to have time for a developing a deep understanding of your area of expertise. Rather be aware that they can expect you to make them understand what they need to know in simple, short statements. If you are a passenger in a car and the driver has not seen a pedestrian, you will shout "watch out for the pedestrian!" and not make a talk about the medical impact of car accidents on pedestrians. Management is the driver and it is your task as a system architect to make them aware of information that is relevant for their decision on how and where to drive.
- Never use a good relation to management for gaining more authority over architecture stakeholders. The system architect's influence in shaping the system of interest comes from the system architect's ability to convince others with well-founded reasoning and not via management dictate. Always solve a conflict with a stakeholder in personal dialog with that stakeholder and never use management as the mediator of your conflicts. Management can expect you to manage your own conflicts.

Table 10.10 describes the good collaboration between system architects and management as a win-win situation.

Chapter 11

Roles

11.1 ROLES

On systems engineering conferences, we have met people who provide their employers with requirements specifications, architecture descriptions, verification strategies, and many more value-adding deliverables. When we looked at their business cards, some were "Digital Signal Processing Engineer," some may have been something like "Head of Mechanical Design Unit F," and others were "System Architect."

No matter what is written on these peoples' business cards, if we meet them on a Systems Engineering conference then it is probably because they are applying thoughts, methods, or processes that are attached to Systems Engineering. If we meet them in a session about system architecture, then they are probably even mentally involved in systems architecting.

In the development of a system, we may ask who currently performs a systems architecting activity. The answer should be independent of this person's position in the organization, but it should be based on the kind of activity that different individuals are currently carrying out. When talking about the activities in system architecting and the skills and competences needed to carry them out then it is possible to describe them independent from the shape of the organization, which makes the

Model-Based System Architecture, First Edition.
Tim Weilkiens, Jesko G. Lamm, Stephan Roth, and Markus Walker.
© 2016 John Wiley & Sons, Inc. Published 2016 by John Wiley & Sons, Inc.

description of systems architecting reusable across different organizations and robust against organizational changes.

To avoid the need to consider different kinds of organization, the notion of a *role* is being used here. A role is an idealized mental representation of a worker or team that has to carry out a certain set of activities. A role usually goes along with a *role description*, defining the objective of the role as well as tasks, responsibilities, and competences that are related to the work to be done. Like an actor can play the role of a president without ever having been elected, a role can be assigned to a person who has not or not yet been given the corresponding position in the organization. This chapter will focus on roles without looking at the organization. Considerations about systems architecting in the context of the organization can be found in Chapter 19.

Persons with the system architect role are central in system architecture. They have to ensure that system elements fit together during system integration and together act as a whole that satisfies the system requirements. They take care that proper architecture and its description are in place and understood in the organization. *System architects* in the terminology of this book are the people having the system architect role. Some companies may give a different name to this role, and often the people filling the role will have other roles in their daily work. This chapter describes what people do and how they behave when they act as a system architect or as a member of a system architecture team, and what skills they need to do so.

11.2   THE SYSTEM ARCHITECT ROLE

11.2.1   Objective

The objective of the system architect role is described in very general terms here. It will be based on the system architect's interaction with stakeholders, as described in Chapter 10.

The system architect ensures that the system architecture of the system of interest is consistent and satisfies the system requirements as well as potential quality criteria.

The system architect ensures that the system architecture description correctly describes the system architecture and that its stakeholder-specific views are understood by the corresponding stakeholders.

The system architect ensures that the system architecture is agreed with the appropriate stakeholders via the system architecture description (key objective: agreed specification of interfaces).

11.2.2   Responsibilities

The system architect is responsible for

- identifying stakeholders, their concerns, and viewpoints to define the views needed to address the concerns;
- creating the system architecture description with appropriate perspectives and views;
- identifying stakeholders who need to review the system architecture description;
- establishing traceability between the system architecture description and the system requirements or use cases and the discipline-specific architectures like software or mechanical architecture;
- conducting reviews with the identified stakeholders and ensuring that the system architecture satisfies the requirements or enables the use cases it is traced to;
- providing stakeholders with expertise about the system architecture and the work on it.

11.2.3   Tasks

The system architect has a multitude of tasks. In order to find out whether a task to be done is one to typically assign to the system architect, one can check if at least two of the following criteria are met:

- The task is multidisciplinary.
- The task addresses a system-level solution.
- The task requires defining or analyzing interfaces in one or more perspectives of the system architecture description, for example:
  - Cross-subsystem interfaces of the physical perspective
  - Interrelationships between functions of the functional perspective
  - Cross-layer interfaces in the layered perspective.

Here is a nonexhaustive list of typical tasks for the system architect:

- Contribute to feasibility studies and effort estimations
- Help project leaders breaking down the work in projects
- Contribute to project planning and verification planning

- Find solutions together with technical experts
- Ensure each interface is agreed with stakeholders on both sides of the interface
- Create the system architecture description
- Publish and communicate the system architecture description
- Explain the system architecture, based on the system architecture description
- Answer stakeholders' questions, based on the system architecture description
- Improve the system architecture
- Assist in interpreting system level test results (e.g., to state a hypothesis in which subsystem to look for the cause of a failure during system verification)

11.2.4 Competences

The system architect has the competence to make architecture decision as long as they do not affect schedules or budget of other stakeholders.

The system architect has the competence to lead negotiations about architecture's decisions in which the stakeholders need to be involved, for example, because their schedules or budget are affected by the decision.

The system architect has the competence to approve the system architecture description.

11.2.5 Required Skills of a System Architect

The main skills of the system architect are abstraction and communication skills. Communication skills are needed because the system architect will need to interact with the stakeholders of the system in order to make the right system architecture and in order to ensure that it is understood by those that need to understand it. The communication skills are so important that most parts of Chapter 20 in this book are about communication. In this chapter, we will therefore not elaborate further on these most important skills of the system architect and will rather go on to explaining the abstraction skills.

Abstraction according to the original Latin roots of the word is about drawing something away. In this case, we draw the essence of a concept out of all the nonidealities of the concept's materialization in a real world. This is shown in Figure 11.1: The system architect in the picture

Figure 11.1. Abstraction skills © 2015 Jakob K., reproduced with permission.

is looking at a relatively complex system, which had to be built in a very complicated way due to whatever constraints. Due to her abstraction skills, the system architect does not think about the system in the complicated way in which it appears. She rather grasps the essential principle of operation, which is "C flows from A to B." The figure has been chosen to be very close to the mentioned literal meaning of "dragging something away." The system architect in the picture drags the principle of operation into her mind by looking at the concrete representation from the system.

Even though abstraction is thus based on the concrete reality, no one prevents us from making an abstraction of a concrete system that has never been built so far. This is the main strength of abstraction in systems architecting: by finding an abstract representation of a system to be built, the system is reduced to its essentials. These can then be taken care of with special focus during the development of the system, even before the first implementation of the system exists.

A system architect with abstraction skills can

- accept descriptions of system behavior that exclude the "pathological cases" and can thus generate simplified descriptions of the system;
- record very complex interrelationships and still see structure in the system.

There is also a related skill, which we would like to call "reverse-abstraction": Once the system architect has derived conclusions from an abstract model of the system, these have to be translated back into the concrete world. In the case of a marketing stakeholder, this could mean as follows: the system architect has to explain the consequences of an architectural constraint on the way the product is perceived by its customer. In the case of an engineering stakeholder by contrast this could mean as follows: the system architect has to explain the consequences of having chosen pattern XYZ on the way the information has to be encoded before being transmitted via a wireless link.

The system architect of course needs more skills than mentioned so far. Some of them will be discussed in Section 11.6. A skill set we do not further elaborate on are leadership skills. These are important for system architects, because leadership can lead to followership, and followership is needed for both ensuring that a defined architecture is being followed and that the way of thinking in systems architecting is being rolled out to those in the organization that need to apply it. Since this is no leadership training book, we direct the interested reader to numerous books that are available about leadership.

11.2.6 Required Skills for Model-Based Systems Architecting

One may think that knowledge of modeling language and modeling tools are the most important skills in model-based systems architecting. Of course these are helpful, but the most important precondition for being successful with model-based systems architecting, in our opinion, are as follows:

- The already mentioned abstraction skills, because all models are abstractions of reality
- Awareness about the "single source of truth" paradigm (see Section 9.11.1) and the ability to create followers for it in the organization
- Understanding of the separation between view and model (see Section 7.7).

11.3 SYSTEM ARCHITECTURE TEAMS

There is often more than one person with the system architect role. Several different system architects can work on the same architecture description, for example, by working on a common architecture model. They should ensure coherence of the system architecture by working in close collaboration. They should work together as a system architecture team, which is a role of its own. The team should ensure that team building happens and team rules and processes are agreed, the team meets often and an agreed agenda exists. The system architecture team links system architects that work in different areas or activities. For example, each of them can be allocated as the system architect in a different feature team (e.g., Larman and Vodde [85]).

Via the system architecture team, the system architects can aim at ensuring the following:

- They follow the same methods and best practices.
- The system architecture stays coherent.
- The way of working with stakeholders is consistent such that systems architecting becomes a predictable and consistent activity for stakeholders, who collaborate with more than one system architect.
- There can be a second opinion on difficult questions.
- More than one system architect understands the system architecture such that a member can be substituted while absent.
- The workload generated by certain activities can be spread to more than one person.
- The system architects can help each other.

There are multiple ways in which system architects can help each other, apart from giving each other assistance on all different kinds of details during daily work:

- People with understanding of different engineering disciplines can join their understanding during the team work, such that multidisciplinary understanding is reached in the team.
- People with different strengths and weaknesses can use their strengths to compensate for others' weaknesses.
- People with different approaches to the same task can find a better approach by choosing the best elements from each of the approaches.

When staffing a system architecture team, one should keep in mind to mix people whose strengths and weaknesses can compensate for each other and whose approaches are different in order to increase the likelihood of finding a good approach for each task. For example, one should mix pragmatic people with highly analytical people. Reasons for this are as follows:

- A team staffed only with analytical people can be trapped in too thorough analysis that does not stop when the solution suffices.
- A team staffed only with pragmatic people can be trapped in producing inconsistencies or chaotic results.

In very small organizations, the total head count limits the number of people that can participate in a system architecture team. But even in larger organizations, there is a limit to the number of persons who can efficiently collaborate as a system architecture team. When we saw winning system architecture teams in real organizations, they were often staffed with exactly six people. We do not have well-founded explanation why a team of six people would be the optimum winning team, however we can think of the following trade-off regarding team size:

- To be as multidisciplinary as possible, the team should consist of people with a background in ideally all the different fields of engineering that are needed for the development of the system of interest.
- In order to still be able to focus on the rest of the team, each team member is potentially limited by the own "seven ± two rule" [96], which means in the "minus two" case that there is a limit of five people to focus on for each team member. Counting also the team member from whose perspective this is counted we come to a total of six team members.

Once a system architecture team has been created, it can start advancing the practice of systems architecting in addition to sharing and doing the operational work.

11.4   SYSTEM ARCHITECTURE STAKEHOLDERS

We have seen typical system architecture stakeholders in Chapter 10. Some of them have dedicated roles in systems architecting. For example, requirements engineers are the ones delivering the requirements input. Others will become relevant in case a procedure states to "involve the stakeholders." For example, there may be a procedure stating that a certain piece of architecture description has to be reviewed by the stakeholders. Which stakeholders to involve is then dependent on the content of this architecture description. If it has focus on production, for example, then the system architect may need to involve production people, whereas verification people may need to be involved in cases in which the test access points of the system are under review. In the descriptions of certain systems architecting activities like the review of a piece of architecture description, it is therefore sensible to use the generic role "stakeholder" instead of a specific role like "production infrastructure engineer."

11.5   RECRUITING SYSTEM ARCHITECTURE PEOPLE

As stated earlier, this chapter is about roles and not about positions or job titles in an organization. When we talk about "recruiting," then we talk about spotting people who are the right ones for the system architect role. Whether they have to be hired first or just found inside the organization does not necessarily matter.

It is an observation that key players in development are often recruited as system architect. This is of course due to the fact that these persons have the overview of at least parts of the system, and creating overview is one of the central tasks of the architect. The question is whether the technical skills that most system architects have are their most important skills. After having stressed that communication and abstraction skills are the most important ones for the system architect, the answer we give is a "no, but ...."

A system architect needs technical skills to interface to technical experts, which will be most comfortable in expressing themselves in technical terms only understood by another expert. A system architect also needs to spread the ideas of new architectural concepts, and therefore needs credibility and in consequence trust among other engineers. Technical excellence is one key in being credible among other technically excellent people. Yet, a system architect who is heavily involved in making system architecture cannot at the same time keep up with the state of the art of the technologies of the system of interest. The system architect will rely on other experts to provide up-to-date technological information, as far as it is relevant to system architecture.

Even more important than the skills are the right talents and the right mindset. Systems architecting needs people with a multidisciplinary mindset and the ability to carry thoughts into different disciplines. These are also known as "T-shaped persons" or "T-shaped individuals" (Stickdorn and Schneider [131] p. 111; Kelley [73]. p. 75), because they have a deep understanding of one area and the willingness to think multidisciplinary, which leads to a graphical visualization in the shape of the letter "T" (the vertical line of the letter represents the deep understanding in one area and the horizontal line represents the multidisciplinary thinking).

Due to the need for systems architects with a good network to the stakeholders, the needed technical understanding and credibility among engineers in different areas, a good place for recruiting system architects is your own organization. Try to find the T-shaped people. Look for those people who are always called by others in case a multidisciplinary problem needs to be solved. Look for those whose absence makes people nervous during such problem solving.

Also look for people who often expose themselves by trying to make things better. Not the ones who can complain about everything and postulate that it can probably be made better. Choose the ones that actually initiate improvements or at least always have a proposal for improvement ready. Leandro Herrero calls these people "simply healthy restless people [...] with a mixture of [...] frustration and at the same time commitment to make things better" ([52], p. 282). He does so in the context of looking for change agents. Actually, one can say that systems architecting requires people with change agent skills, for example, because ensuring that stakeholders stick to an interface agreement may mean that they have to *change* the solution idea toward one that complies with the interface definition, but also because establishing a multidisciplinary mindset sometimes requires the initiation of a mindset *change*.

Finding the right people is more important than looking for people with the right position in the company. For example, people with a good sense for project management are not necessarily good architects, because they have a focus on timing rather than on the right concept and a common understanding (see Figure 10.2)—and actually one should try to avoid assigning the project leader role and system architect role to the same person, because the roles have to act in different directions (Figure 10.4). Still, also a system architect needs the ability to finalize tasks according to schedule instead of getting lost in details.

System architects are ambassadors for the system architecture in the different areas of engineering. It is thus good to look for "unofficial leadership" in these areas: A development engineer who is often invited for technical discussions based on wishes from the other participants is a good candidate for becoming a system architect. The fact that other team members wish to have this person in the room when discussing certain problems can be a sign that this person has facilitation and mediation skills and the ability to bring a solution forward—thus typical skills of a system architect. Of course, there may also be persons who are omnipresent because they have succeeded making themselves "indispensible" in the organization for instance by not sharing their knowledge. These persons are also invited to many activities for obvious reasons. They are bad candidates for becoming system architects, because knowledge sharing is essential in architecting a system.

Recruiting system architects within the own organization of course also brings some challenges:

- The people with the needed skills are often regarded as "key players" and their superiors may claim that it is impossible to handover their work to others.
- Also when having the system architect role, these people may still get involved in their old work. Care has to be taken that this will not always win the battle for priority.

In situations in which such challenges are met, the following thoughts may be helpful: When a key player is removed from a team of development engineers and replaced by someone else, the organization becomes less dependent on "heroes" (i.e., the key players without whom nothing works). Furthermore, these persons will feel a career jump by being nominated as system architect and might take this as a reason to stay some more years with the company. Thus the know-how of the former "heroes" stays in-house and can be handed over to others gradually.

When recruiting system architects, also think of the system architecture team they will be part of: which talents or skills are missing? Are there more team members with analytical approaches or more pragmatic people? Try to recruit architects that bring the missing talents, skills, and approaches into the team.

11.6   TALENT DEVELOPMENT FOR SYSTEM ARCHITECTS

Each of us can go to a dancing course, but some will have to learn dancing by practicing hard, others will just seem to dance by themselves. The latter ones are of great talent for dancing. Assuming we have recruited people of great talent for systems architecting as described before, will they in future architect the system optimally based on their talent? Probably not. Just like a talented dancer can still learn from the right teacher, system architects also need to keep learning, partly from being trained in methods and tools, partly from gaining experience on the job.

When we have recruited a new system architect, we have most likely found a "T-shaped person": Apart from an existing background in an own engineering discipline, there is a talent for multidisciplinary thinking and as a consequence a broad knowledge. The system architect will now be trained in system architecture methods, processes, and tools, but also in other systems engineering practices, for example, in the disciplines of the important systems engineering stakeholders like requirements engineering and verification. Furthermore, the system architect will learn on the job while doing coordination work between different engineering disciplines. This will not only cover systems architecting practices, but also the specialties of different technologies and disciplines that are related to the subsystems. The system architect will gain different depths of understanding related to different subsystems.

Persons who have proven enough professional excellence in the system architect role may get tasks that aim at improving the systems architecting discipline itself rather than the system of interest. These people will thus have less contact with the operational engineering business than some of their colleagues. As a consequence, they will no longer be close to the carriers of up-to-date knowledge in different engineering disciplines and thus their own knowledge of these disciplines will become outdated.

If we summarize this development from a T-shaped person to an experienced system architect who is able to improve the systems architecting approaches of the own organization, then we see a development like it is shown in Figure 11.2: The figure shows the described development of the architect in three snapshots from the left-hand side to the right-hand side in diagonal upward direction. The vertical dimension in each of these snapshots symbolizes the depth of technical understanding in different technologies or subsystem-related disciplines. The horizontal dimension that is pointing away from the reader symbolizes the depth of understanding for different systems engineering disciplines, and the remaining horizontal dimension allows for showing different disciplines next to each other.

The left snapshot in Figure 11.2 shows the situation in which a system architect is often recruited: We see the "T" shape of a T-shaped person, which in general of course implies more than can be shown in the figure. Comprehensive knowledge in systems engineering will usually be missing in this stage, unless the person has been to a systems engineering education or has prior experience in systems engineering. In these cases, Figure 11.2 is not applicable.

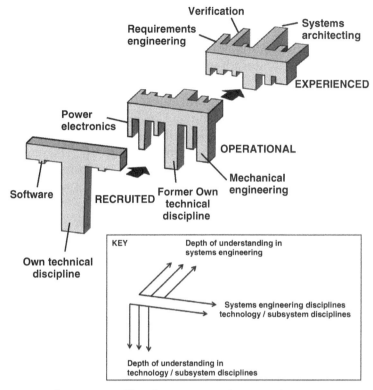

Figure 11.2. Changing profile of a system architect with growing experience, based on the assumption that there has not been any systems engineering career or education upfront.

The middle snapshot in Figure 11.2 resembles very much a profile shown by Maier and Rechtin [92], as it will be discussed soon. It shows the situation of a system architect who has become operational in the systems architecting work: knowledge in different technologies, engineering and systems engineering disciplines is present. The knowledge in the former own technical discipline has degraded. This does not necessarily mean that the system architect has forgotten anything learned during an earlier job, it rather relates to the fact that technologies and the corresponding engineering approaches evolve and change fast such that everyone who is not following the related technical discipline with full dedication will no longer be up to date. For the system architect, this is the case for most engineering disciplines, due to a conflict of the required dedication to with the new focus on systems architecting. The loss of proficiency in the own area of technology is an important aspect to consider for engineers who like to become system architect. Only when accepting this loss, they can gain proficiency in systems architecting with full dedication. As already mentioned, Maier and Rechtin show a quite similar snapshot in their book "The Art of Systems Architecting" ([92], pp. 8–9). They trace it back to a lecture by Bob Spinrad at the University of Southern California in 1987. While they show the *"required* depth of understanding" in the different subsystem disciplines and omit the systems engineering dimension, we show the actually obtained depth, comprising systems engineering. The reason for us not to show the required depth like Maier and Rechtin was the preceding explanation about the loss of proficiency in the original engineering discipline. It would not be visible if we only sketched the proficiency actually required for being a system architect. Maier and Rechtin can use their representation to explain another important aspect regarding the depth of understanding: it is not always required to understand each subsystem in detail, however the system architect has to "dive in" in some cases and find out the very details of a task at hand. For this reason, systems architects will gain deeper understanding in different disciplines.

The rightmost snapshot in Figure 11.2 shows the profile of the experienced system architect. Detailed subsystem understanding has now been replaced by understanding for the different systems engineering disciplines, first and foremost of course systems architecting.

Note that Figure 11.2 shows a view on a mental model, which is like every model an imperfect representation of reality. It is due to that imperfection of our model and due to the fact that the different facets of human talents and human personality are unlimited that we expect we could find excellent system architects all over the world whose career or current profile has nothing in common with the profiles shown in Figure 11.2. We have constructed the model to warn about the potential loss of technical proficiency in the system architect's former own area of expertise and to motivate the need for trainings in systems engineering methods and tools. However, we highly discourage the use of this model for talent spotting or for measuring the skill level of a system architect.

So which are the trainings we can use to accompany the even more important learning by doing of the system architect? Here is a nonexhaustive list for your inspiration:

- Trainings in processes (not only in system architecture but also in requirements engineering, verification and validation, change and

configuration management, and in other systems engineering disciplines)
- Trainings in systems architecting methods (e.g., in the FAS method we will describe in Chapter 14)
- Trainings in a modeling language (e.g., in SysML)
- Trainings in the tools used for supporting the above-mentioned methods and processes as well as the modeling language, so for example a SysML modeling tool
- Communication and Presentation trainings.

Since we see on the job experience as the most important aspect of talent development for systems architects, we recommend to make a systematic planning not only of the training in methods and processes, but about their application in real-world projects. In the best case, both happens simultaneously, such that training contents become tangible in the daily work immediately.

Chapter **12**

Processes

12.1 THE SYSTEMS ARCHITECTING PROCESSES

The systems architecting process is the core process for the system architect. Table 12.1 shows an example how the inputs and outputs of the process as well as the contributing roles could be defined. The major output is of course the system architecture description (see Chapter 6). But also the learnings from carrying out the systems architecting work should be captured during the work and should be phrased as heuristics (see Section 7.10.1), for example, in a best practice document. This leads to a living document to be updated and used across projects.

The actual process for producing the system architecture description consists of doing the architecting and validating or reviewing[1] and approving its output.

A very simple explanation of systems architecting is to translate requirements into an allocation of functionality and performance to subsystems. However, in many cases one may need to collect the most architecturally relevant requirements before the other ones, because:

- The requirements may not yet be completely in place at the time at which the very basic architecture of a system needs to be decided on

[1] See Section 12.1.3 for a discussion of the terminology.

Model-Based System Architecture, First Edition.
Tim Weilkiens, Jesko G. Lamm, Stephan Roth, and Markus Walker.
© 2016 John Wiley & Sons, Inc. Published 2016 by John Wiley & Sons, Inc.

- Requirements may change during a project, for example, based on new market inputs.
- The system architecture may have to be reused across multiple projects, whereas a requirements document often has one single project as its scope.
- Maintainability and scalability may be needed in the future, during other phases of the life cycle.

This is a reason for inputs like the quality criteria. We will elaborate on this when talking about quality criteria and quality requirements further below.

The system architect has to ensure that the system architecture considers each life cycle stage of the system and its evolution. In our experience, most system architectures live longer than they were intended to live. Therefore, the architect has to consider how to make the architecture endurable, which may be one of the quality criteria we see as an input in Table 12.1. Here is a discussion of all the inputs in the table:

- The system context provides the scope for the system requirements, the use cases, and the system architecture. See Section 8.2.

TABLE 12.1 Examples of Inputs, Outputs, and Contributors for the Systems Architecting Process

Input to the systems architecting process	• System context • Requirements • Use cases • Quality criteria • Heuristics, patterns, and best practices • Roadmaps • Domain knowledge
Outputs from the systems architecting process	• Architecture description • Updated heuristics and best practices
Contributing roles	• System architect (drives the process) • Stakeholders (provide input and consume outputs)

- Requirements and use cases specify the expected functions, behavior, and performance of the system. The system architecture has to realize the satisfaction of the corresponding expectations.

- Quality criteria give guidance what criteria to optimize for (e.g., flexibility vs. simplicity). These can be obtained in close collaboration between system architects and their stakeholders (see also Section 10.10). The system architect should ensure that the stakeholders agree on the quality criteria. Ideally the requirements engineering process provides quality requirements for each of the quality criteria. However, we recommend that the system architect ensures that quality requirements are in place for each of these, in close cooperation with requirements engineering. The high importance of the quality requirements results from their meaning regarding the life cycle of the system architecture: while most requirements aim at a certain product portfolio that is known at the time of the requirements elicitation, quality requirements drive architectural decisions that are difficult to change during the whole life cycle of the system architecture. Since fundamental architectural decisions are often valid for more than one product, it is important to look ahead of the products that are usually in focus of the requirements engineering processes. The system architect should ensure that the quality criteria are obtained from the stakeholders. It may not be optimal to have the stakeholders phrase quality criteria alone, because dependent on their background they may miss the necessary understanding of the architecture impact. It is therefore very important that close communication between the system architect and the stakeholders happens. The system architect should ensure that it is clear to the stakeholders which questions need to be answered by defining the quality criteria. Quality criteria drive the definition of quality requirements, which can be exploited during architecture assessment (see Chapter 18). For example, a quality criterion could be modifiability.

- Heuristics, patterns and best practices can help solving particular challenges of the given business. Typically, they are maintained throughout many projects and updated with the learnings from each project.

- Roadmaps can give an idea about the scalability requirements.

- Domain knowledge complements requirements in creating understanding about what has to be developed.

The INCOSE systems engineering handbook [56] defines a process for systems architecting. It names inputs and outputs partly different than Table 12.1. Among others, it mentions the "System Element Requirements." There is a model-based way of producing these in the context of model-based systems architecting. See Figure 9.20. In our nomenclature, the creation of them also involves the requirements engineer (see Section 10.2).

After having made a very general description of the systems architecting process, we will now show some examples of the process steps, first in a very general fashion and then with some relations to the contents of other chapters in this book.

12.1.1   Example of Generic Process Steps

Here are typical steps of the systems architecting process, in a very generic phrasing:

- Identify stakeholders and their concerns
- Identify perspectives (these may be given by the standardized architecture framework used in the organization)
- Identify views for the stakeholders
- Identify quality requirements, that is, the ones describing characteristics that have been identified as quality criteria. Further above we chose modifiability as an example quality criterion. A related quality requirement could be: "A new color variant of the system must be ready for sales 3 months after the start of development."
- Evaluate architecture alternatives based on the quality requirements and decide which one to choose. For example, to stick with the quality requirement of fast color change, a modular architecture may be chosen with a strict separation of colored housing parts and those parts that contribute to core functionality of the system (according to the pattern "separate stable from unstable parts" in Section 7.5).
- Model the Architecture by information of different views and mapping the different views
- Produce the architecture description by generating the views
- Validate or review and approve the architecture description.

12.1.2 Example of Concrete Process Steps

Here is a proposal on how to use concepts from this book in the context of the systems architecting process (it is highly recommended that you tailor it to your personal needs by removing some steps or adding some):

- Use stakeholders from Chapter 10 as a default initialization of the stakeholder list, and assess whether the given project needs additional stakeholders
- From Chapter 9, select the system context, the functional perspective and physical perspectives as well as traceability perspectives
- Ensure that the system context description is available
- Ensure that the base architecture meets the quality requirements according to the quality criteria, based on architecture evaluation (see Chapter 18)
- Use the FAS method (Chapter 14) to fill the functional perspective, based on use cases from the input of the process
- Make functional views for the different stakeholders with their functions of interest and the functions having functional interfaces with them (see Chapter 9)
- Do the functional-to-physical mapping together with the stakeholders, based on the requirements
  - Formal alternative
    * Find alternative mappings
    * Use trade studies/decision theory (Section 17.4) to choose the optimum alternative. Use the requirements as selection criteria and ensure that particularly the nonfunctional requirements get enough focus
  - Informal alternative
    * Assemble a team with representatives of all physical blocks that are potentially involved in the mapping [82]. In some approaches, this is called a feature team.
    * Give the requirements to this team and ask them to come up with a good functional-to-physical mapping and a reasoning why they chose the mapping.
- Generate the architecture description from the produced models, views, decisions and reasoning (e.g., using the template from Section 6.3.1)

- Validate or review and approve the system architecture description according to Section 12.1.3.

12.1.3 Validation, Review and Approval in a Model-Based Environment

Validation is a term with multiple definitions in different standards for different industries. Sometimes, it refers to an activity that confirms if an architecture description document satisfies the requirements, for example, by means of a review. Sometimes, it refers to an activity that confirms if a sample of the product meets the stakeholder requirements when being used in its operational environments by sample user. Sometimes, it can be used for both or even for different kinds of activities. Since we like to address readers in different industries, we are not using the term any further in this section.

This section is about the confirmation that the contents of the architecture description are correct. There are two major aspects to this:

- The system architecture as described by the architecture description satisfies the system requirements.
- The system architecture as described by the architecture description is technically sound.
- The architecture description is consistent in itself.

Again, there are multiple versions of processes for achieving this. Let us make a very simple example of a process, in order to use it for discussing the details of reviews and approval in model-based systems architecting:

1. The system architect identifies the stakeholders that need to be involved in the review.
2. The system architect conducts a review with the identified stakeholders.
3. Potential findings from the review are addressed and the process is restarted, if necessary.
4. As soon as it is appropriate to confirm that the contents of the architecture description are correct, the system architect declares the architecture description *approved*.
5. The above process is documented, for example, by listing persons involved and their roles, dates of reviews, actions taken to address

review comments and measures taken to verify that they have been done. It may also be appropriate to state which reviewers were involved in the creation of the architecture description under review and which ones were independent reviewers.

The above process may be too simple to be applied in industries or around systems with regulatory requirements toward processes or products. We recommend to the reader to retrieve information about the applicable regulations before adopting any steps of the above process.

In a document-based environment one can simply follow the above process if it complies to the applicable regulations. It can be conducted on the basis of an architecture description document, and then the outcome can be documented inside the document under review, for example:

"On 2015-02-20, James Blonned (assigned image processing engineer), Roger Mouhr (researcher applied physics, independent reviewer) and Singeon Smyth (assigned optical engineer) reviewed the document, found it to be correct, consistent and in compliance with the stated system requirements. As a consequence, Eyeaen Flaemeeng (assigned system architect) approved this document on 2015-02-20."

This is of course an example, and dependent on the applicable processes or even regulations the statement may look different and an electronic or handwritten signature might be required, for example, by means of a validated tool.

In a model-based environment there is in the first place no document that one can review, approve, or sign. This has consequences for the approval process but also for regulated processes that may require a signature. For this book, we cannot consider the regulatory aspect since it is very much dependent on the context and on the applicable regulations. We will therefore continue with the simplistic assumption that there are no regulations to take into account. We will consider how to ensure that information which is retrieved from the model can be provided with a maturity state indicating whether the appropriate stakeholders have reviewed and approved it.

During the 2014 annual workshop of GfSE (the German chapter of INCOSE), one of the authors (J.G. Lamm) and Professor J. Abulawi discussed the approval of model-based architecture descriptions [1]. J.G. Lamm proposed to export selected views as pictures and to approve them by signing them electronically or on paper. The notion would then be that views and model elements that are invisible in this selection should be considered nonexisting. Model-Based Document Generation (see Section 6.3) was seen as the means of supporting the export of selected views in the form of a document. This approach can still be considered a fallback solution in case other approaches seem impractical. However, the result from the discussion [1] was that a different approach could be chosen for proper model-based proceeding. It will be briefly presented in the following.

Based on the separation of view and model (see Section 7.7), let us assume that the model is the actual information carrier, whereas views on it simply lead to visualizations of the information. This means that the model itself should carry the information whether it has been reviewed and approved. However, since the stakeholders will access information from the model via stakeholder-specific views, these will be used for preparing the information under review.

How to get from a review via views to an approval of the model? First it has to be ensured that each view contains the model elements that are relevant for the corresponding stakeholders. Rules should be stated, defining the criteria for a model element to appear on a view. For example, if a functional architecture model is under review by a feature team working on one functional block, the given functional block together with all functional blocks that have connections with it via ports could be in scope. In this example, the view for the given feature team would thus show blocks that are also in scope of a similar view for other feature teams. The same effect can occur in other perspectives. In Section 9.3, we showed different views from the physical perspective, which have a high amount of overlap regarding the model elements they show (Figures 9.6 and 9.7).

As we saw in the example, a model element can be in multiple views, which have to be reviewed by different groups of stakeholders. This means that one single review activity cannot come to the conclusion that a model element is approved. It can only come to the conclusion that the element is approved in the context of the view that was used for the review. We introduce the maturity-state *preapproved* for views. It indicates that the review conducted on the given view has concluded to approve what has been reviewed.

Based on the preapproval, automation can be added to automatically identify a model element as approved if all the views in which it appears have been preapproved. The pseudo-code below illustrates this:

```
boolean isApproved (model element m)
   for each v in all views
    if v contains m
            if v is not preapproved
                return false
            end if
       end if
    end for each
    return true
```

If it is required to produce approved visualizations, one may need to verify if a certain view is approved in order to automatically generate a maturity state to associate to the view.

The information given on a view is approved, if all the visible model elements are approved as defined above. The following piece of pseudo-code illustrates this:

```
boolean isApproved (view v)
      for each m in all model elements
          if m in v
             if not isApproved ( m )
                  return false
             end if
          end if
      end for each
      return true
```

If the whole system architecture model should be approved by means of the above approach, then it will have to be ensured that each model element is contained in at least one view.

It is now up to the reader to consider if the above solution is practical. At least it requires automation and it also has a drawback: model elements and views that were already in the approved state lose this state as soon as a new view is created, because it can be assumed that the view will not be preapproved on creation, thus all model elements of the view can no longer be approved, and in consequence the same goes for all views showing those model elements.

12.2   CHANGE AND CONFIGURATION MANAGEMENT PROCESSES

The change and configuration management process in the scope of systems architecting ensures that

- versions of the system architecture description or even of separately versioned parts of it are put under configuration management;
- changes on approved system architecture descriptions are carried out with the needed focus on impact analysis and reapproval of the changed artifacts.

The discussion of the collaboration with the configuration manager in Chapter 10 provides more details on this topic. The system architect should consider taking the configuration manager role in case there is no configuration manager.

12.3   OTHER PROCESSES INVOLVING THE SYSTEM ARCHITECT

There are several processes in other disciplines that involve the system architect. They are the ones ensuring that the necessary cooperation between system architects and stakeholders takes place. One example is the requirements engineering process in which the system architect will for example assess the feasibility of satisfying a requirements with the envisioned or already given system architecture.

Chapter 10 provides a discussion of typical stakeholders in system architecture. The processes around them are typical ones that could also involve the system architect. Since this book is focused on systems architecting, those processes will not be discussed in detail.

Chapter **13**

Agile Approaches

Being agile can have different meanings, and it does not necessarily mean to drop everything in order to do the task that was shouted loudest or with highest priority. On the contrary, typical agile approaches aim at finishing one thing in order to be ready to think about what is important to do next.

The authors are convinced that agile approaches have many advantages on the waterfall-based proceeding that was in place during the time when they started emerging. We see organizations that have adopted agile practices in parts of their development entities or even on a broader basis. So far, we have only seen very few that are fully agile in systems engineering and also present this to the outside world. This may be due to the fact that the introduction of agile practices on an organizational level is a challenging change process.

When agile approaches are implemented partly and differently in different entities of the organization, the system architect will be affected by the differences, being a multidisciplinary agent who has to collaborate with multiple entities inside the organization. This chapter aims at showing the real-world situation of system architects today, who cannot necessarily live agile approaches according to the textbooks.

Model-Based System Architecture, First Edition.
Tim Weilkiens, Jesko G. Lamm, Stephan Roth, and Markus Walker.
© 2016 John Wiley & Sons, Inc. Published 2016 by John Wiley & Sons, Inc.

We start with a brief summary of important events in the history of agile development. We then show the potential contrast between visions and reality of agile practices that the system architect has to be prepared for.

13.1 THE HISTORY OF ITERATIVE-INCREMENTAL AND AGILE DEVELOPMENT

In 1986, the article "The New New Product Development Game" by Takeuchi and Nonaka [133] proposed product development with self-organizing teams that are working according to a "unique dynamic or rhythm" and see learning as an important practice. The whole article uses metaphors from the rugby game to underline both the aspect of moving forward as a team and interacting with others in a "scrum." The scrum metaphor has survived, in the name of the *scrum* approach of agile development (e.g., see Ref. [85]).

In 1998, The *spiral model* by Boehm was published [16]. Targeted at software development, it proposed an alternative to the traditional sequential development processes Boehm called "the waterfall model." The alternative was an iterative approach that would cycle through the different stages of a traditional approach multiple times, each time adding an increment to the product. Boehm called it an "evolutionary development model." It is known as the spiral model, because it was visualized by drawing development iterations as a spiral around the center of a coordinate system. Figure 13.1 is a very rough sketch of such an incremental approach without showing a spiral and with strong simplifications compared to Boehm's model. It shows that each iteration is composed of

- doing engineering activities (which here should be assumed to include among others requirements engineering, systems architecting, and implementation);
- verifying and validating the work products;
- evaluating the achievement and preparing for the next iteration.

In 2001, the agile "Manifesto" for software development was published on http://agilemanifesto.org/. It emphasizes priority of certain values over traditional techniques in software projects. For example, it assigns high value to interaction and individuals, and it values collaboration with the customer more than the negotiation of contracts.

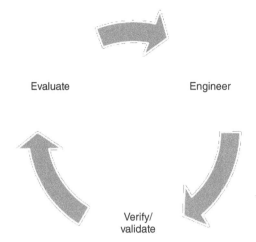

Figure 13.1. Typical iteration cycle of agile development.

All the mentioned approaches have inspired modern agile development. Even though some were initially targeted at software, also the systems community today knows approaches based on iterative–incremental development cycles that each result in a "potentially shippable product increment" (e.g., Larmann and Vodde [85]), which is not supposed to be shipped under all circumstances but reminds the developers to finish one task before they start the next one. In this context, it is maybe worth to mention that the initial "scrum" approach by Takeuchi and Nonaka [133] was inspired by typical systems like cameras and copiers.

Another approach that comes from systems development rather than software development is the "lean" approach. Larmann and Vodde call the book we cited in the previous paragraph "Practices for Scaling Lean & Agile Development," which shows that there are similarities between agile approaches and lean ones. In systems engineering, there is a working group of INCOSE that focuses on lean approaches. It is the INCOSE Lean Systems Engineering Working Group. Some of their work results have been summarized in Oppenheim's book "Lean for Systems Engineering with Lean Enablers for Systems Engineering" [110].

Agile practices have changed the approach of categorizing priorities with few discrete states like "low," "medium," and "high" into one distinct priority for each task, which allows for defining a precise order in which tasks have to be carried out. This again supports the notion of finishing one task first before finishing the next one. This would maybe not be possible if several tasks all had the priority level "high."

Modern agile practices are thus about empowering self-organizing teams to work on getting things done one after the other, in an iterative-incremental approach with frequent evaluations that allow for changes and learning to occur. It is this ability to change often based on evaluating what has been done that makes these practices agile.

Last but not the least, we would like to recommend the book "Agile Modeling" by Amber [8]. It addresses modelers that would like to improve their effectiveness during modeling by applying a similar focus on getting work done as developers who apply agile development practices. Even though the book addresses software developers, its recommendations are suitable for modeling in systems architecting. It is full with hand-drawn sketches and white-board prototypes that remind us of the card technique for applying the FAS method in a workshop that we have described in Section 14.9.

13.2 SYSTEM ARCHITECTS IN AN AGILE ENVIRONMENT

System architects are dependent on involvement of stakeholders from different engineering disciplines in order to get architecture increments and other deliverables done. They are thus also dependent on the development approaches of their stakeholders. We see in many organizations that the different areas of engineering are working with different degrees of implementing agile techniques. For the system architect this sometimes results in a mixture of agile and traditional approaches among the stakeholders and sometimes in a mixture of different agile approaches. Even if all stakeholders that are involved in a task are working according to the same agile approach, some practical issues may occur for the system architect, for example, different cycle lengths or starting points of the cycles through Boehm's spiral in different departments of the organization.

As a consequence, there is a challenge for the system architect, who is not working in an organization with a holistic agile approach (like it is e.g., described by Larmann and Vodde [85]). This challenge should be met by being flexible in applying potential own agile practices of a system architecture team, but also by knowing and proactively coordinating the different work approaches of the different stakeholders. The system architects should know which of their stakeholders follow an agile approach and which ones follow a traditional one, and for those following an agile approach, the system architect should know when a new cycle through Boehm's spiral starts and how to inject requests for work packages to be placed on the cycle.

Ideally, the system architects in an organization seek the dialog with those people responsible for the implementation of agile approaches and make them aware of the need to produce deliverables across different entities in the organization.

Chapter 14

The FAS Method

The functional architecture is often mentioned in publications about systems engineering or in the context of real system projects. On the second view, you will realize that different terms are used like logical architecture, logical view, or functional view. And different artifacts are part of those architectures: set of functions, flow models, or functional models for simulation purposes. However, they all share the same notion: a technology-independent, function-oriented description of the system.

Jesko Lamm and Tim Weilkiens have observed a lack of concrete common methods for functional architectures in particular in the context of MBSE. Some years ago they described the FAS (Functional Architectures for Systems) method [83]. It was not a complete new method, but more a putting together of already existing puzzle pieces. The FAS method is a practice-proven method based on common MBSE practices. We describe FAS in this chapter and start with our view on the terminology and motivation of functional architectures.

Finally, we shine a light on different aspects of functional architectures like tool support, nonfunctional requirements or the role of technology in a technology-independent description.

Parts of this chapter are based on our article "Method for Deriving Functional Architectures from Use Cases" [84]. We omit the citation of

Model-Based System Architecture, First Edition.
Tim Weilkiens, Jesko G. Lamm, Stephan Roth, and Markus Walker.
© 2016 John Wiley & Sons, Inc. Published 2016 by John Wiley & Sons, Inc.

the article at each statement that is taken unchanged or updated from that article.

14.1 MOTIVATION

Functions are the essential core of a system. Their results are finally provided to the user of the system. They are the primary important features and any other aspect is secondary or depends on the functions. Therefore, it is useful to make them explicit artifacts in the development process of a system.

The functional architecture enables the description of functions in an intuitive way that fits into the toolset of system architects. It is a description of systems independent of their technology in a block-oriented structure. System architects have their strength thinking in blocks and their dependencies and not in functional flows, which is more the strength of the requirements engineers.

A descriptive example for the separation of function and technology is the photography. A function for taking photos was present already in an old-fashioned mechanical Camera Obscura. The Camera Obscura is no longer a state-of-the-art technology in photography, but modern cameras still need the related function of enabling the photographer to take photos. Of course, the decomposed functions on the more detailed levels are different between the Camera Obscura and modern high-end cameras. These functions depend on the selected technology. Technologies introduce new functions that are directly bound to those technologies. See Section 14.12 about the relationship of technology and technology-independent functional architectures.

This example shows: describing products by their functions leads to concepts with higher lifetime than an approach that depends on a certain technology. And the lifetime of technologies is steadily decreasing and new technical components require an update of the functional deployment. Especially, a shift of a functional allocation from one engineering discipline to another, for example, from electrical to software, could be best analyzed in the holistic view of the functional architecture.

We observe an increasing tendency that functions are deployed on many physical components and that a physical component provides more than one function (Figure 14.1). That advances the need for a clear understanding of the function structure. In the past, it was sufficient for an engineer to focus on the physical parts of the system. Simply spoken, one function was allocated to one physical part and

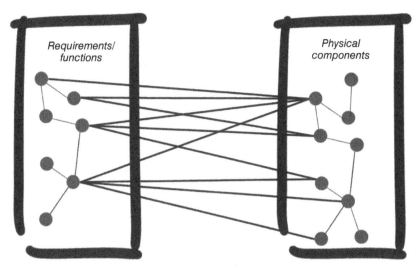

Figure 14.1. Functional-to-physical mapping.

one physical part realizes one function. The related functions of the physical parts were obvious and well covered. That's different if many functions are realized by the physical part and a function is realized by many physical parts. Which are these functions? Which ones are the most important ones? What is really optimized if the engineer improves a physical part? The functional architecture gives the answers to these questions.

A functional view also enables a deeper insight into the system [2, 53]. By using model-based functional architecture descriptions, system architects can model their understanding of the system of interest without the complication of simultaneously designing a concrete technical solution. Functional architectures are a well-established concept (e.g., Refs [11, 19, 53]).

There is a gap between the requirements and the physical architectures. An organizational gap: Requirements engineering and system architecture are different disciplines with different roles and typically different people behind the roles. Quite often the people are organized in different departments and work at different locations. A cultural gap: Requirements engineers and system architects have often different educational backgrounds and different mindsets.

That leads to misunderstandings in communication. Although you can model a mapping relationships (e.g., satisfy) between the architecture and the requirements, there is a gap and the other side gets easily out of the view from the perspective of the requirements or architecture.

The functional architecture does not completely fills the gap, but is a good bridge pier to narrow the gap. It is closely related to the requirements and at the same time part of the convenient environment of the systems architect toolset in the form of a block-oriented view. It is a place where both disciplines meet.

According to the INCOSE Systems Engineering Handbook, the functional architecture is the foundation for the system architecture through allocation of the functions to system parts [56]. The systems engineering manual of the Federal Aviation Administration describes the functional analysis as an activity "that significantly improves innovation, synthesis of design, requirements development, and product integration [99].

Since the functional architecture is an essential and comprehensive description of the system functions, it is a suitable artifact to be used for functional safety assessment. One of the first steps of a safety assessment is to identify the appropriate functions, for example Ref. [59], that could be found in the functional architecture.

14.2 FUNCTIONAL ARCHITECTURES FOR SYSTEMS

Model-based functional architecture descriptions model systems independent of their target technology, by means of functional elements that transform modeled information (signals, data), materials, force, or energy [115, 139]. Functional elements have different names in the literature, for example, *functional element* [150], *function* [75], or *system block* [14]. Hence, it is necessary to define terms that relate to functional views of the system like we use them in this book. This has been done in Table 14.1. Definitions of more common architecting terms like architecture or architecture description are listed in Chapters 4 and 6.

In addition to the definitions in Table 14.1, Figure 14.2 shows the terms defined in a domain model with further information. The left side of the figure shows the SysML model elements that represent elements of the functional architecture or system requirements and use case analysis artifacts.

Functional elements can be decomposed into subelements. The corresponding task is called *functional decomposition* and its result, the *function structure*. Function structures are hierarchical. Their topmost level of hierarchy is closely related to use cases. Not all approaches of a functional system description have this topmost level of hierarchy. The concept of functional decomposition can be found in a variety of sources ([113]: 170; [33]: 145–146; [11, 129, 115]: 66; [115]: 199). These sources

TABLE 14.1 Terminology

Terminology	Term Definition
Function	Input/output relationship [113, 115] of information (signals, data), materials, force or energy within the system of interest, or a model thereof.
Functional element	Abstract system element that defines a relation between at least one input and at least one output by means of a function.
Functional decomposition	Decomposing functions into subelements. The result is a function structure.
Functional group	Set of strongly related use case activities.
Functional interface	Set of inputs and outputs of a functional element.
Functional part	Usage of a functional element in a functional architecture.
Function structure	The hierarchical structure that results from decomposing functions into subfunctions and from decomposing functional elements into subelements (Term based on Ref. [113], but with a modified definition).
Functional architecture	Architecture based on functional elements, functional interfaces and architecture decisions.
Use case	Description of usage of a system, in terms of achieving a goal that is realized by providing a set of services and values to selected human actors, external systems, or stakeholders [106].
Use case activity	Description of one or more behavioral elements from the coordinated sequencing of actions that take place when a use case is instantiated (based on Refs [26, 106]).

differ in background, and not all of them have a systems engineering context. We interpret functional decomposition for systems in general and thus do not specifically emphasize one field of engineering, as it is done for example in Ref. [113], where there is a strong focus on mechanical design.

We distinguish between functional, nonfunctional, and constraint requirements. Most of the functional requirements are refined by use cases with their related use case activities (see Section 8.4). The nonfunctional requirements specify the quality of the functions or other nonfunctional requests like legal conformance. The constraint

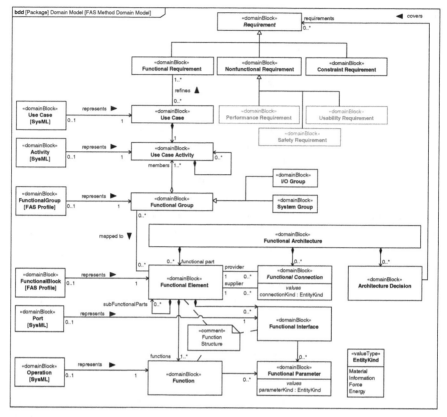

Figure 14.2. Domain model for functional architectures.

requirements are rules or facts that must be observed and not satisfied by the system. For example, the base architecture is a set of constraint requirements (Section 7.2). Typically, you have a more fine-grained classification of the requirements like usability, performance, or business requirements. All are subtypes of functional, nonfunctional, or constraint requirements. Figure 14.2 shows some examples like performance or safety requirements.

The functional interface specifies inputs and outputs of a functional element. The grouping of inputs and outputs to functional interfaces is done by the system architect based on criteria that are relevant for the architecture.

The functional architecture consists of functional elements, their connections, and architecture decisions. For instance, a nonfunctional performance requirements could lead to a decision of a system architect how to pattern the functional architecture.

How to model functional architectures with SysML is described in Section 14.5. You can replace SysML with other notations, for example text written in natural language, spreadsheet documents, or your proprietary graphical notation. Functional architectures are independent of SysML although they perfectly fit together. Next we give a SysML independent description of the FAS method although we use SysML to visualize some concepts of FAS.

14.3   THE FAS METHOD

Jesko Lamm and Tim Weilkiens described a method for deriving a functional architecture description directly from use cases in an intuitive and traceable way [84]. They called it the FAS method. The tasks are performed by the system architects in collaboration with the requirements engineers using a clear interface and handover between both roles. The requirements and use cases are represented mainly in text and flow diagrams that are well suited for requirements engineers. While the functional architecture of the FAS method is a structural description in a block-oriented representation that is well suited for system architects.

The FAS method was first published in a paper for the German systems engineering conference TdSE in 2010 [83]. The first international publication of the FAS method was the paper "Method for Deriving Functional Architectures from Use Cases" in the Systems Engineering Journal [84]. This chapter is an extended version of that paper. Since its initial publication, the method has been used in various industry and research projects (e.g., Refs [31, 78, 84, 152]).

The FAS method derives functional architectures in a block-oriented form from use cases, via the grouping of use case activities. The method is based on standard techniques in systems engineering like the identification of the system context and subsequent use case analysis (e.g., Refs [145, 147]). A brief description of these tasks can be found in Chapter 6. They are not part of the FAS method, but deliver work products as input for the method.

A key step of the FAS method is the grouping of use case activities into functional groups. The grouping avoids developing different physical system assemblies for the same functionality. This part of the FAS method is the one that most requires the expertise of the architect and can also be supported with heuristics to be presented in Section 14.4.

In Section 8.3, you find some results of the use case analysis of the VMT example system that are input work products for the FAS method. We are interested in the set of use case activities. They are grouped according to use cases, that is, the grouping criterion is the usage of the system from the actors' perspective. The functional grouping of the FAS method applies a different grouping criterion. It is the cohesion of the use case activities, that is, use case activities that belongs to the same subject are in the same group. Figure 14.3 shows the different views. While the use case view is important for the requirements analysis and to incorporate the nontechnical stakeholders of the system, the functional view is important with respect to the system architecture to implement a system that satisfies the requirements.

Figure 14.4 shows an example of a functional grouping. We put all use case activities in a matrix where the activities are the columns and the functional groups are the rows. That matrix gives a good overview of the grouping and is a handy tool to do the grouping of the activities to functional groups.

Use case activities are grouped according to the cohesion principle with regard to their subject (Section 7.15). For example, in Figure 14.4 the functional group "Tour Management". The group includes all use case activities that deals with museum tours. Members are for instance the use case activities "Confirm end of tour" or "Select tour."

Use case activities with a relation to external interfaces are grouped in special "I/O" groups. That follows the heuristic "One functional group takes the functions that are related to system actors" described in Section 14.4.2. The system architect could then put her focus on the non-I/O functions that are typically more essential for the system.

Top-level root use case activities are sometimes too comprehensive to be completely assigned to one of the functional groups and are assigned to a special top-level functional group called "Top-Level Functional Group."

A good practice for the structure of the functional groups is a subject-specific structure at the first level and the separation of the I/O functions on the next level. The I/O functions are again separated according to the related system actors (Figure 14.5). One functional group could contain other functional groups and each functional group except the root group is a member of exactly one other functional group (tree structure). The best practice for your project depends on many criteria like team organization, focus on usability requirements, and more. Finally, the structure of the functions is a decision of the system architect.

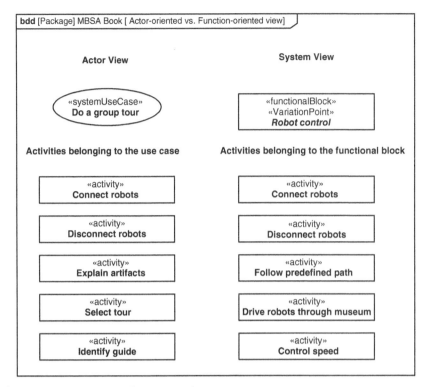

Figure 14.3. Views on the system functions.

Functional groups are mapped to functional elements of same name. The transition from functional groups to functional elements decouples the life cycle of use cases from the one of the functional architecture, facilitating the implementation of change control across systems analysis and system architecture. For the first step, each functional group is mapped to exactly one functional element. Later, the functional groups could be different than the functional elements. For instance, the system architect refines a functional element and split it up to three functional elements. All three functional elements relate to the same functional group.

Functional elements and their functional connections can then be modeled in block-oriented form, as illustrated in Figure 14.6. The connection specifies that the output of functions from one functional element is an input of functions of the appropriate connected functional element. The connections are linked via functional interfaces with the functional elements. The functional interfaces describe the input and output entities of the functional element. A functional element could

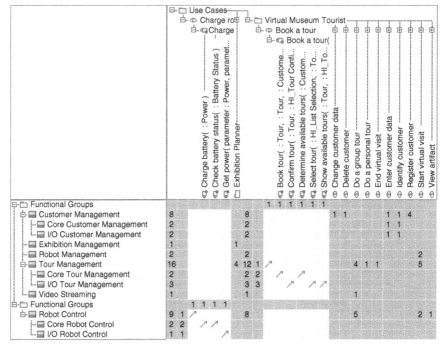

Figure 14.4. Matrix functional grouping.

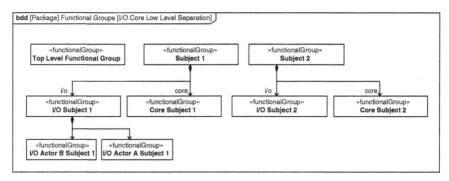

Figure 14.5. Functional groups with hierarchies.

have more than one functional interface to group inputs and outputs according to aspects that are important for the system architect. The inputs and outputs of a functional element can be derived from the inputs and outputs of the appropriate use case activities.

The functional elements together with functional connections and architecture decisions are the functional architecture.

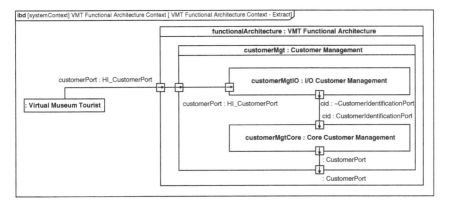

Figure 14.6. Part of the VMT functional architecture.

14.4 FAS HEURISTICS

Architecture combines art and technology and cannot be automatically created. Although some automated tasks can support the architect. The main work is based on the experience of the architect and guided by architecting principles (Chapter 7). In addition to automation also heuristics can support the architect [92]. In the following, we provide heuristics to create groups of use case activities. The grouping aims at functional groups that are "as independent as possible; that is, […] with low external complexity and high internal complexity" ([92]: 28). This is a known concept in classical architecture [4] and has been discussed by Baylin [11] regarding the functional modeling of systems. According to Baylin, functions should be clustered according to "functional cohesion" ([11]: 32), that is, with the aim of keeping elements with related or same objectives together. See also Section 7.3 about the cohesion and coupling principle. For supporting the architect Lamm, Lohberg and Weilkiens [82, 84] have found heuristics that aim at functional cohesion. They will be presented in the following.

14.4.1 Abstract and Secondary Use Cases Define a Functional Group

The use case model already reveals potential functional groups. An abstract use case represents commonalities between several concrete use cases and indicates a cohesive grouping. Its activities are good candidates to be assigned to a functional group.

A secondary use case represents behavior that is shared by several use cases. Typically, it represents a cohesive set of functions and is a good candidate for a functional group. The modeling of abstract and secondary use cases is covered by Weilkiens [147, 145]. We give a brief introduction on use case analysis in Section 8.3.

14.4.2   One Functional Group Takes the Functions that are Related to System Actors

Functions having a direct relationship with system actors are part of the system's input/output logic. Often, they only have little in common with the actual system functions that do the processing of the inputs and produce the outputs and are a cohesive set of functions on its own. In that case, they are good candidates for a separate functional group.

14.4.3   Function Calls Imply Cohesion

 Functions call other functions that often relate to a similar topic, resulting in a network of call relationships. Clusters in that network are potential functional groups. They can be derived from the use case activities.

14.4.4   Functions that Share Data can be Grouped

It can be assumed that two functions belong to closely related domains if the kind of output of one of them is the other's input. This is more applicable if the kind of object is not a common data type, but special and related to the domain. This connection can easily be found in the object flow of the use case activities. The so-called activity trees of SysML can facilitate the assessment of common data (Figure 14.7). See also Section 9.9.1 for more information about activity trees.

14.4.5   Use Grouping Criteria of Existing Groups

A system is rarely developed completely from scratch, but it is usually based on an existing system. The outline of existing system documentation from prior or similar systems can indicate possible ways of grouping functions. Ideally, interviews with the system developers should be made to find out if the grouping was useful in practice. This way, known structures will be created and team members will find them intuitive to use. However, grouping criteria have to be reassessed with caution: they can be of technical rather than conceptual nature. A grouping based on technical constraints is not desirable, because it will lead to a functional architecture that contains implicit technological decisions, making it more difficult to find alternative solution scenarios.

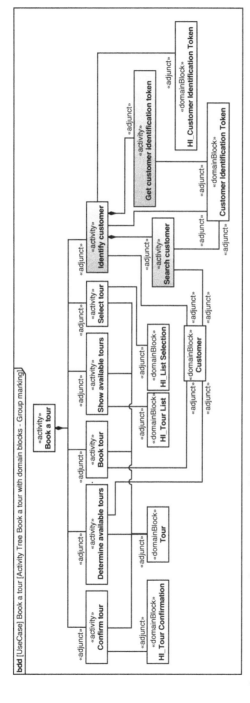

Figure 14.7. Example activity tree of the VMT.

14.4.6 Reduce the Number of Functional Groups that Include Functional Variation Points

A variation point marks an element in the model that could be realized in different variants of the system. See Chapter 15 for more details about the modeling of variants. Reducing the number of functional groups that include functional variation points means to group those functions in as few as possible functional groups. The heuristics follows the architecture principle to separate stable from instable parts (Section 7.5). More concrete G. Schuh states that variant full assemblies should be separated from assemblies with low variability [125].

Remember that all these are heuristics and no rules. They can support and guide the system architect, but not take over the decision for a functional group.

14.5 FAS WITH SYSML

While the FAS method is independent from any language or tool, we will show its application using the OMG Systems Modeling Language (OMG SysML™) [105], because this language is well suited for implementing the FAS method. We recommend to use SysML for the following reasons:

- SysML is an international standard. It is well known, mature, and fits in most collaborative tool landscapes.
- SysML provides all necessary model elements for the FAS method.
- SysML is also suitable for the adjacent models of a functional architecture like requirements and the physical (logical and product) architecture models.

We've previously introduced the FAS method already with the SysML notation. Now we'll have a deeper look on the SysML model elements used for the FAS method.

14.5.1 Identifying Functional Groups

The functional architecture, as we introduce it here, belongs to the structural view on the system and does not include system behavior. As a consequence, the control flows of use case activities are less important in the FAS method, whereas the object flows are of major importance. The object flows connect the inputs and outputs of use case activities.

SysML provides a structural view on activities that hides the control flow. It is the activity tree (also called "function tree" in the literature (e.g., [147, 145])) depicted in a block definition diagram, in which each node represents an activity. Figure 14.7 shows an activity tree for the use case activity "Book a tour." Activities with a common object are highlighted by means of a different fill color of the use case activity (according to the heuristic "functions that share data" from Section 14.4).

The tree structure expresses the functions' call hierarchy; that means: a node calls its child nodes. This does not make the activity tree a functional decomposition where the node owns its child nodes. The lines connecting the activities in the tree are the SysML part association relationships. The properties typed by an activity are SysML adjunct properties. They constrain the values of the properties to be determined from the appropriate call behavior action.

The roots of activity trees are the activities that have a one-to-one relationship to the use cases. Each use case owns exactly one root use case activity that specifies the overall behavior of the use case. These activities have the same name as the use cases they relate to, as an exception to the naming convention for activities that the root activities are named from the perspective of the system actors and not from the system perspective like all other activities that are based on system functions.

It has to be noted here that even though the activity tree view as shown in Figure 14.7 is suitable to identify functional groups, the practical application in modeling tools may sometimes require a different representation, because activity trees may grow very large for systems of realistic complexity. A replacement of the view is the matrix view of relationships that is provided by most SysML modeling tools. The columns represent the activities and eventually actions, the rows the functional groups, and the marker in the cells represent the grouping relationship (Figure 14.3).

The activity trees are based on the SysML activity and not the action model element. In SysML, these are different concepts [147, 145]. The SysML activity is the whole behavior depicted by an activity diagram, and the SysML action is an atomic part of an activity. Functional decomposition is modeled with call behavior actions that call an activity, which acts like a decomposition.

Despite the functional architecture, it is the best practice in use case modeling to model each use case step with an activity to define it and a call behavior action to describe its usage. It is another example of the definition/usage/runtime pattern (Section 7.4).

If you need other action kinds than the call behavior action, for instance for simulation purposes, be aware that these actions are not considered by the activity trees with SysML like we describe it here. However, you can tailor the usage of SysML for FAS to include any kind of actions. In that case, you can't use the tool of activity trees to identify functional groups. But they are just a tool and not mandatory for the FAS method. Instead of activity trees, you can use matrices to visualize and create the relationship between the activities or actions and the functional groups. A functional group can include activities as well as actions and functional requirements that are not refined by use cases. It could be any element that represents a function definition.

However, be careful not to mix up activities and actions. A call behavior action is a usage of an activity and not a definition of a function. In that case, the called activity is an element of the functional group and not the call behavior action. Same for the call operation action that is a usage of a block operation. It is the operation that defines the function and is part of the group. An opaque action is a definition and usage of a function at the same time and could be an element of a functional group. Actions to receive or send signals are also potential candidates to be a member of a functional group.

In SysML, functional group is depicted by a SysML block with stereotype «functionalGroup» from the FAS profile.

The membership relationship between a functional group and the activities is the SysML trace relationship. The functional group is the source and an activity the target of the trace relationship.

A functional group can contain another functional group. The hierarchy relationship between functional groups is modeled with the composite relationship. Figure 14.8 shows the relationships in a SysML block definition diagram. In addition, the lower level functional groups could be moved into the namespace of the owning functional group. In modeling tools, the namespace structure is often used in the model browser and matrix representations and helps to get an overview of the functional groups (Figure 14.9).

14.5.2 Modeling the Function Structure

In SysML, the function structure is represented in block definition diagrams. Per functional group, one functional element is modeled as a SysML block with a stereotype «functionalBlock» defined in the FAS profile. To keep the traceability between the functional block and the

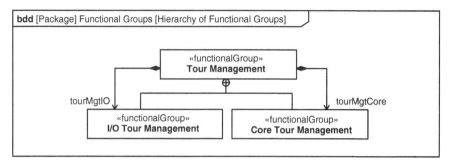

Figure 14.8. Hierarchy relationship between functional groups.

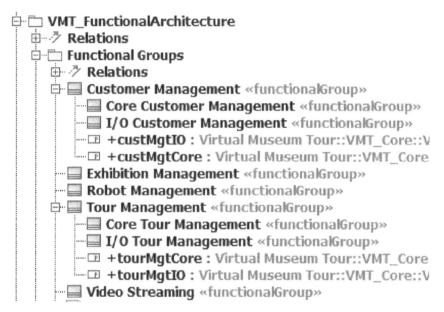

Figure 14.9. Namespace containment of functional groups.

functional group a trace relationship is modeled from the functional block to the functional group.

Again that relationship could be best shown and managed in a matrix. Figure 14.10 shows the trace relationship between the functional block and functional groups of the VMT. At a glance, you can spot that the functional block "Robot Position Tracker" and "Robot Utilization Tracker" have no trace relationships to functional groups. They were added later by the system architect to refine the functional block "Robot Management." There is no appropriate functional group and the traceability to the use case activities and requirements is

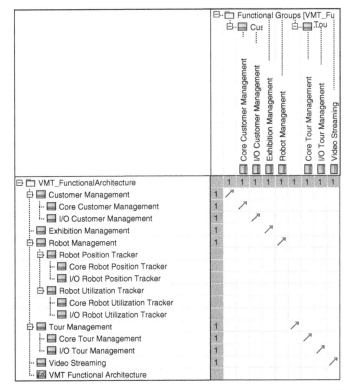

Figure 14.10. Traceability between functional blocks and functional groups.

established by the enclosing functional block "Robot Management" that has a trace relationship to a functional group.

The operations of the functional blocks could model the actual function, that is, the input/output relationship between objects that are the block's input and output via its ports. As a rule of thumb, each operation of a functional block should match a function of the corresponding functional group. The functions of a functional group are the use case activities of the group. You get the traceability from the operation to the use case activity by sharing the same name or more formally, but with more effort with a trace relationship from the operation to the activity element. The latter could again be managed with a relationship matrix.

Note that with modeling this relationship, you lose the decoupling of the functional architecture from the use case model that you get by the indirect relationship via the functional groups. Figure 14.11 depicts the relationship between the operation "Select robot()" and

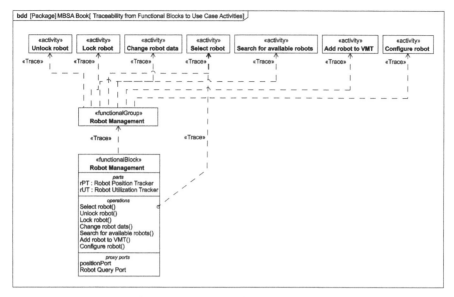

Figure 14.11. Traceability from functional blocks to use case activities.

the appropriate activity. The topology of the diagram clearly shows that the decoupling of the functional group is bypassed by the trace relationship between the operation and the activity.

If you don't need that level of detail in the functional architecture, you can skip the modeling of the operations. If you need that level of detail, you connect the operations of the functional blocks directly with the appropriate activities and you can discard the functional groups. In such a scenario, the functional groups are just a tool to support the derivation of the functional architecture.

Functional blocks can be described as parts of other functional blocks via the composite relationship. This can be used to model the decomposition of functional elements into subelements. In the first step, the composition hierarchy is a copy of the composition hierarchy of the functional groups. Like described for the functional groups above, the composition hierarchy of the functional blocks could be mirrored by the namespace containment to structure the model.

Figure 14.12 shows parts of the function structure of the VMT. The root element is a special functional block that represents the complete functional architecture. It has the stereotypes «functionalBlock» and «system» applied.

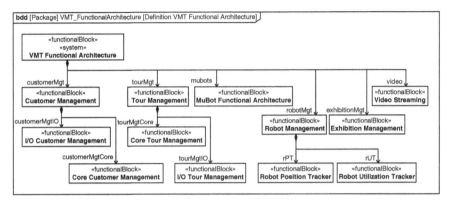

Figure 14.12. Function structure of the VMT.

14.5.3   Modeling the Functional Architecture

Functional block parts whose functions output is input of functions of other functional block parts are connected in the internal block diagram via proxy ports with flow properties for the input and output data. If the proxy port does not delegate the flows to or from the internal parts of the functional block part it should be marked as a behavioral port, that is, it represents behavior of the owning block.

Connections between ports can be looked up partly at the object flows of activity models; in practice, however, internal block diagrams can describe certain matter with a more clear visualization and provide a better overview than activity diagrams — particularly in cases of objects flowing between functional elements resulting from different branches of the activity trees.

The internal block diagram depicts the usage of the functional blocks and the connections based on the input and output of the functions that are encapsulated by the blocks. That structure is the functional architecture description.

Figure 14.13 shows one possible functional architecture description of the VMT in a SysML internal block diagram. In this case, the system is shown within its context, which allows us to also display its external interfaces and the actors they are connected with. Hence, the interrelation between system functions and external systems or system users can be visualized.

Figure 14.14 shows the definition of the port types of the functional blocks. The port types are not yet complete. The *HI_CustomerPort* is a user interface. «userInterface» is a stereotype of the SYSMOD profile to mark port types that describes the direct interface to a human

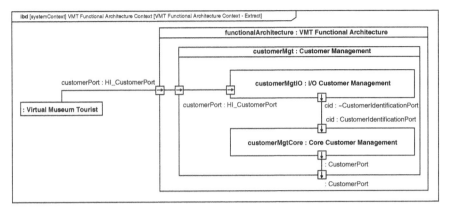

Figure 14.13. Example functional architecture of the VMT.

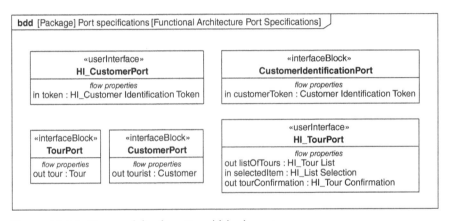

Figure 14.14. Types of the functional block ports.

user. It is a special SysML interface block to type proxy ports (Section A.2.2). In Figure 14.13, you can see the proxy port "customerPort" typed with the user interface "HI_CustomerPort" at the functional part "customerMgtIO." There is a path of connected proxy ports also typed "HI_CustomerPort" to provide the interface also at the enclosing blocks to finally connect the system actor Virtual Museum Tourist.

Another port at the functional block "I/O Customer Management" specifies the interface to handover the data to internal customer management core functions. The port type "CustomerIdentification-Port" is an interface block. The tilde symbol before the name marks the part as conjugated port, that is, the direction of the flow properties are reversed and also provided features of the interface block are requested and vice versa. That way the part perfectly fits to the connected port at the functional block "Core Customer Management," where the port of the same type "CustomerIdentificationPort" is not conjugated.

In SysML, the separation of the definition of blocks in block definition diagrams and their usage in internal block diagrams fits to the definition/usage/runtime pattern described in Section 7.4.

The functional architecture could be assessed using assessment methods described in Chapter 18 and architecture principles described in Chapter 7 like the cohesion and coupling principle.

14.6 MODELING TOOL SUPPORT

The FAS method interlinks use case analysis with the creation of a functional architecture. This enables traceability between requirements, use cases, and functional elements if there is a modeling repository or tool chain that allows for linking both. Once a modeling tool has been chosen, it is usually possible to script certain modeling steps in order to automate the user's work and thus make modeling-intense work less error-prone. Indeed, the following parts of the FAS process from Figure 14.15 are well suited for being automated [78, 84]:

- The creation of initial functional groups in the step "Identify Functional Groups"
- The creation of initial functional blocks and their interfaces in the step "Model Functional Architecture."

By the time of writing this book, the open source community supports the FAS method with add-ons that are offered for download for two different modeling tools (www.fas-method.org, March 2015), providing some of the possible automation.

In the following, one typical approach of the automation will be described. For simplicity, the term FAS engine will be used for the

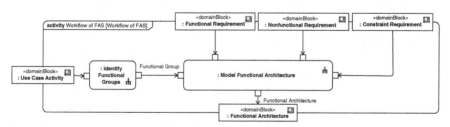

Figure 14.15. FAS process.

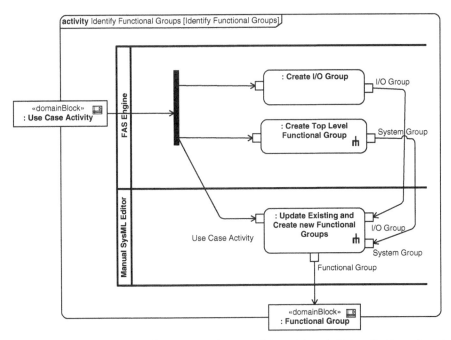

Figure 14.16. Manual and automated steps of the FAS task "Identify Functional Groups".

modeling tool add-on that supports the FAS method. To illustrate the split of the FAS method's steps between manual work in the SysML editor and automated actions of the FAS engine, Figure 14.16 shows the detailing of the task "Identify Functional Groups" and Figure 14.17 shows the detailing of the task "Model Functional Architecture" from Figure 14.15. The split between manual and automatic tasks is modeled by partitions.

14.6.1 Creating Initial Functional Groups

The heuristic "One functional group takes the functions that are related to system actors" can be supported already during the use case analysis phase: the call behavior actions of the use case activities are split into activity partitions representing the core functionality of the system of interest and those relevant for interfacing it to its actors. The activity partition of the latter kind can be marked with the stereotype «I/O» from the FAS profile [78].

Figure 14.18 shows an example of the use case activity *Book a tour* with activity partitions. On the first view, there seems to be a mistake in

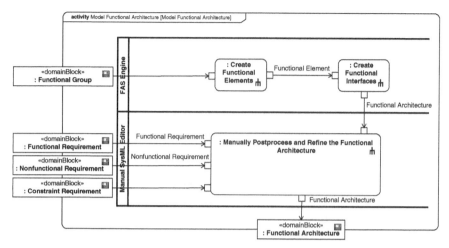

Figure 14.17. Manual and automated steps of the FAS task "Model Functional Architecture".

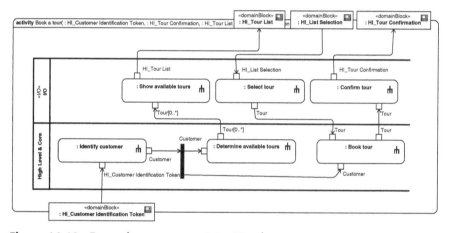

Figure 14.18. Example use case activity "Book a Tour"

Figure 14.18. The input parameter "HI_Customer Identification Token" flows directly in an action of the core partition and not in an action of the I/O partition. On the second view, you will see that the activity "Identify customer" is created to avoid redundancy. The behavior is reused in other activities. The activity has its own I/O partition (Figure 14.19). There the input parameter "HI_Customer Identification Token" is input of an action of the I/O partition. On the use case level, "Identify customer" is modeled as a secondary use case (see also 8.3).

In the step of grouping activities, the FAS engine can automatically group all activities that are called from actions in the I/O partition into

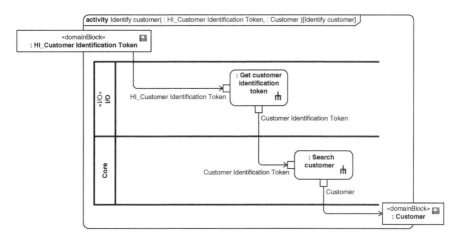

Figure 14.19. Example use case activity "Identify Customer".

one functional group "I/O" ("Create I/O group" in Figure 14.16). In addition, a FAS engine could determine potential conceptual mistakes. An activity should not be called from actions in an I/O partition and from actions in a non-I/O partition; input and output parameters of the activity that are connected with actors should flow to or from an action of an I/O partition.

Furthermore, the FAS engine can group all nodes of the activity tree, except the leaves, into one "Top-Level Functional Group" (second action in Figure 14.16). This ensures that top-level activities in the tree are assigned to a group, and that the functional block for representing the whole system of interest itself will be created later and traced to that group.

Figure 14.20 shows a typical result of the automatic initialization of functional groups. The leftmost column displays the created functional groups, which are modeled as SysML blocks with stereotype «functionalGroup». The two mentioned groups "I/O" and "Top-Level Functional Group" have been created, and activities from our example system have been assigned to the functional groups by means of trace relationships, pointing toward these activities. The membership is represented by small lines in the cells of the functional grouping matrix.

14.6.2 Changing and Adding Functional Groups

The architect will now manually edit the functional grouping (last step according to Figure 14.16). Additional functional groups will be created

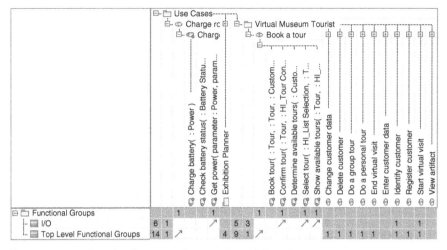

Figure 14.20. Result of automatic creation of function groups.

manually according to the heuristics mentioned further above in Section 14.4, and the architect will have to assign the activities to them, until the assignment of activities to groups represents a disjoint decomposition of the set of activities. A typical result of such work is shown in Figure 14.3.

14.6.3   Creating Functional Blocks and their Interfaces

Functional blocks are created automatically by the FAS engine (first step in Figure 14.17): For each functional group, one functional block of same name is created. The functional block is linked to the functional group it originated from, by means of a trace relationship. All blocks are then made parts of the functional architecture by a composition relationship from the functional block that represents the whole functional architecture of the system to all the other functional blocks. Furthermore, the composition hierarchy of functional groups could be created automatically between the functional blocks.

Interfaces in functional architecture (ports and their connections in Figure 14.13) can be created with tool support (second step in Figure 14.17), because interfaces between functional blocks depend on the object flows in the underlying use case activities. The FAS engine can verify by itself if there are object flows between activities behind different functional blocks and if there are, it can suggest creating an interface between the corresponding blocks.

The architect may now modify the created functional architecture (last step in Figure 14.17). The result of such work is shown in Figure 14.13: While most blocks and ports are unchanged compared to a first version of the functional architecture that had been synthesized with the FAS engine, some additional connections and ports toward system actors have been modeled. In this step, the architect can also create subelements of functional elements, now considering not only functional requirements but also nonfunctional requirements.

14.7 MAPPING OF A FUNCTIONAL ARCHITECTURE TO A PHYSICAL ARCHITECTURE

A functional architecture itself cannot be implemented. Therefore, a physical solution providing the identified functions is needed before the architecture can be realized in a system. This solution comes with a structured view, which we call physical architecture description. We differentiate two special physical architectures: the logical architecture and the product architecture.

A full-fledged system model has a base, a functional, a logical, and a product architecture. The functional architecture is allocated to the logical architecture and the product architecture is specialized from the logical architecture (Figure 14.21).

Definitions

The *physical architecture* is an architecture that represents the physical elements of a system and their relationships.

The *logical architecture* is a special physical architecture that covers the technical concepts and principles of the system under development [64].

The *product architecture* is a special physical architecture that covers the concrete technical implementation of the system.

In practice, there is often only one physical architecture that has aspects of the logical as well as the product architecture, that is, some blocks represents technical concepts while other blocks represent concrete technical assemblies. A strict separation of the two architecture kinds is useful for instance to get a logical architecture for reuse in other system development projects.

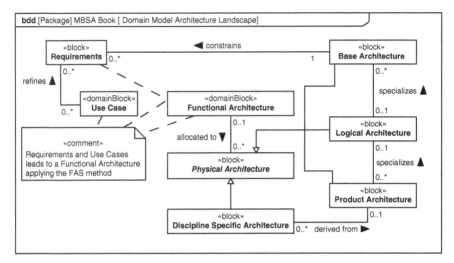

Figure 14.21. Domain model of architecture landscape.

Possible procedures for realizing functions in a physical system have been described in the literature ([53]: 397; [14, 75, 139, 115]: 203). System architects are interested in the allocation of elements in physical architecture from functional elements (Figure 14.22). See Section 9.9.1 for more information about the functional-to-physical mapping.

In SysML, the physical architectures are modeled with block definition and internal block diagrams similar to the modeling of the functional architecture whereby the blocks now represent physical entities of the system like software, electrical, or mechanical parts. SysML provides an allocated relationship to model allocations from functional to physical blocks. That relationship could be best modeled and viewed with a matrix representation (Figure 14.23).

For example, the functional part property "coreCustomerMgt" from Figure 14.13 can be allocated to a physical part property "vmtServer-Application" of a physical architecture of the VMT – meaning that the "coreCustomerMgt" functionality is realized with an application that runs on the VMT server (Figure 14.22). If the mapping is valid in any context, you can also allocate the type of the functional part to the type of the physical part, that is, the functional block to the physical block. The allocation between the properties is only valid in the contexts that define the properties.

One functional architecture can map to different physical architectures. The functional architecture still remains the same. For instance in trade-off studies to assess several physical architectures

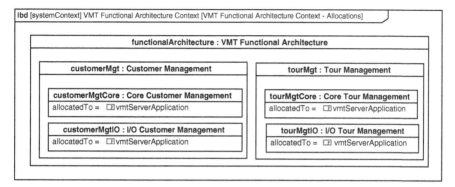

Figure 14.22. Allocation of functional parts to physical parts.

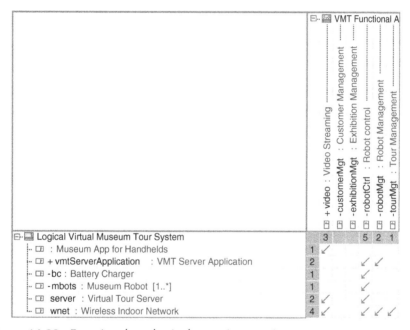

Figure 14.23. Functional to physical mapping matrix.

for the same functional architecture or for product families that share the same functional architecture.

The mapping of functions to physical components requires the experience of the architect. In most cases, system projects do not invent complete new technologies but improve existing systems with partial new components. Therefore, the mapping of most functions to physical components is well known. However, the holistic deployment is rarely documented and prevents optimization strategies. We have observed that system architects tend to have the right gut feeling with regard to finding appropriate functional–physical mappings.

14.8   EXPERIENCES WITH THE FAS METHOD

After its first publication in Ref. [83], the FAS method has been used in multiple industry projects (e.g., Refs [31, 78, 84, 152]). For the practical application of the FAS method in an industrial project, the procedure according to Figure 14.15 has to be mapped to the processes of the project and the organization. The different analysis and architecting tasks involved in the FAS method are allocated to existing roles. Typically, use case analysis will be the task of a system analyst or requirements engineer, whereas the creation of the functional architecture will be a system architect's task.

Use cases are modeled in close collaboration between the requirements engineer and the stakeholders of the system. During that phase, it has been observed by the authors that stakeholders can be overwhelmed by the rather technical representation of detailed activity diagrams. Therefore, we prefer to make a nonformal version of these diagrams that focus on the requirements, before doing a more formal refinement.

The step of separating I/O actions from core actions is a step, which we consider as a good candidate for being carried out in close collaboration between requirements engineers and architects. One of the authors observed that the transformation between activity diagrams from use case analysis and block diagrams of the functional architecture facilitates the communication between requirements engineers and architects: while analysts seem to have their strengths in explaining the expected system functionality with activity diagrams, architects seem to perform best when assessing block diagrams. The FAS method enables both requirements engineers and architects to work in "their" respective world and synchronize each others' findings via architecture generation by means of the FAS method.

In one case that was observed during an industrial project; the architect immediately spotted a concept error after having synthesized the functional architecture with a FAS engine. This concept error had not been found before though the same architect had reviewed the corresponding activity diagrams. The error found in the functional architecture could be easily traced back to the activity model from which it originated, by means of the trace relationships that had been created by the functional grouping and by the architecture creation. The error was then corrected in the activity diagram and the functional architecture was again synthesized. Such fast iterations of architecture synthesis and resynthesis may only be efficient if a FAS engine is available for the used modeling tool.

The paper [31] discusses experiences with the FAS method in different real-world projects. All projects concluded that the FAS method was worthwhile. Most of the projects systems have many nonhuman actors. Although the FAS method is based on use case analysis that is more known for interactions with human users, it is also applicable for networked systems as well. Further observations reported in the paper state that the FAS method costs not much effort, but leads to a valuable benefit and that the FAS method could also be well applied by less experienced people.

14.9 FAS WORKSHOPS

In a FAS workshop, requirements engineers and system architects together create a first version of a functional architecture or a larger update of an existing one. A perfect setting of a FAS workshop is 4–8 participants, an even mix of requirements engineers and system architects, and a timeframe of one day. We do not recommend to use a modeling tool in a workshop environment with more than three people. It is hardly possible to incorporate everyone in an efficient process in such a scenario. Instead, we propose a card technique to elaborate the functional architecture in an efficient way with a group of people.

The procedure of the workshop includes the following steps. Some steps may be skipped if the information is already there. For instance if the workshop should not create an initial functional architecture, but a large update of an existing architecture. In that case, you should consider a brief presentation of the existing information for the workshop participants instead.

- Give a presentation for a brief overview of the FAS method and the action items of this workshop.
- Identify the system context and draw the context diagram on a flip chart. The chart should be visible in the workshop location at any time.
- Identify use cases of the system based on the requirements and the system context.

- Write each identified use case with a unique id number on a single card and pin the card on a wall (Figure 14.24).
- Describe the use case activities and their input and output data. Write each activity on a card with the input data on the left side and the output data on the right side. Each activity gets a unique sub-id according to the use case, for example, "8.5" for the fifth activity in use case #8 (see Figure 14.25). Pin the cards on the wall below the corresponding use case cards. Mind that the order of activity cards does not necessarily need to be the order of execution since it is not important for the functional architecture.
- Find the functional groups with the help of the heuristics from Section 14.4. Write each functional group on a card and pin it on the wall. Again write a unique ID for the functional group on each of the cards. Put the grouped activities below the functional group card. The activity IDs on the cards ensure the traceability to the associated use case.
- Finally, you can sketch a functional architecture as an internal block diagram on a white board or flip chart (Figure 14.26). Each functional group is mapped to one part in the functional architecture. It is not necessary to explicitly define the functional blocks in a block definition diagram during the workshop. They

Figure 14.24. Example of use case cards.

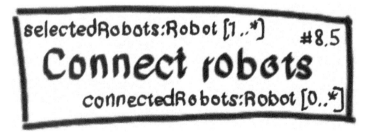

Figure 14.25. Example of a activity card.

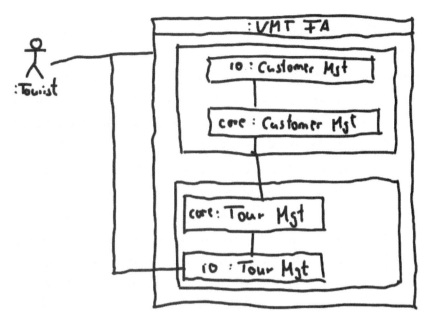

Figure 14.26. Sketch of a functional architecture.

are implicitly defined by their usage in the internal block diagram. The type of the parts, that is, the functional block, should have the same name as the corresponding functional group. In addition you could reference the unique ID of the functional group to be able to change the name without losing the traceability. Connect the parts by analyzing the inputs and outputs of the included functions. Use ports at the parts for the connection points and give them proper names. They are the functional interfaces. If it is important, you can further describe the flow properties that is specified by the functional interface on a separate flip chart. Typically, this task can be done outside of the workshop by the system architects. You don't need the whole team—requirements engineers and system architects—for that.

- Discuss the architecture. Is it conforming to your architecture principles? Seems it proper according to your functional and nonfunctional requirements?
- Take photos of the workshop results and assign an action to a system modeler to incorporate the results into the system model.
- Ask the workshop participants for feedback on the updated system model with the functional architecture.

14.10   NONFUNCTIONAL REQUIREMENTS AND THE FUNCTIONAL ARCHITECTURE

The functional architecture is derived from functional requirements via use cases. That way, most functional requirements are covered well by the functional architecture. See Section 14.11 about how to increase the coverage of functional requirements by the functional architecture.

But what about the nonfunctional requirements? Some of those requirements are directly related with functional requirements. They call for the quality of the functions, for instance the duration or resources needed by a function. The requirements engineer elaborates a matrix that represents the relationships between functional and nonfunctional requirements (Figure 14.27).

Since the functional requirements could be traced to the functional architecture, there is also a traceability from the functional architecture to the nonfunctional requirements (Figure 14.28).

Some of those nonfunctional requirements could be directly incorporated in the functional architecture. Figure 14.29 shows an example of a constraint to satisfy a nonfunctional requirement in a functional

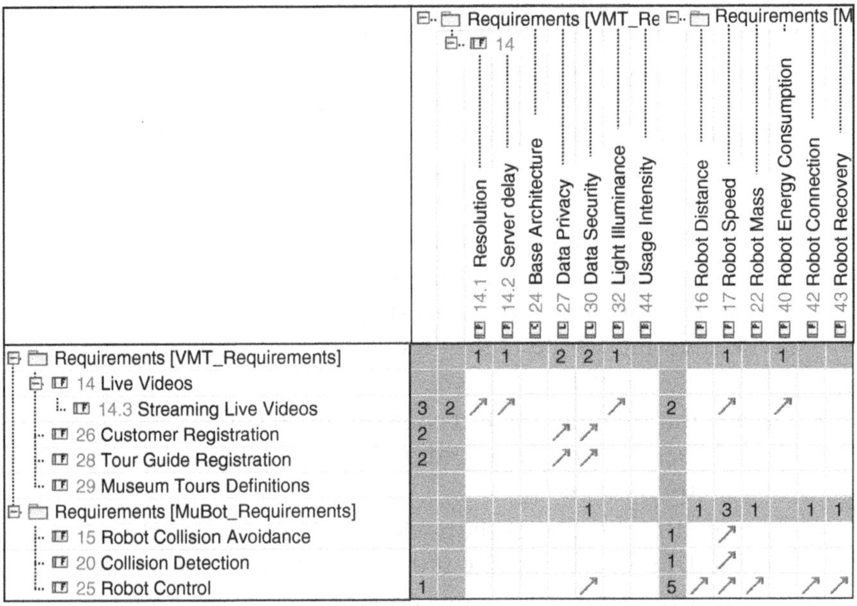

Figure 14.27. Relationship of functional and Nonfunctional requirements.

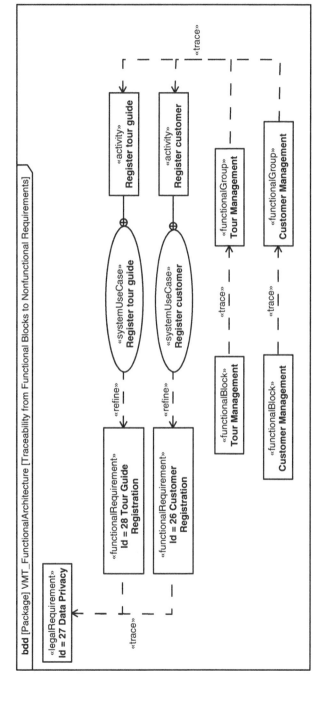

Figure 14.28. Traceability from functional architecture to nonfunctional requirements.

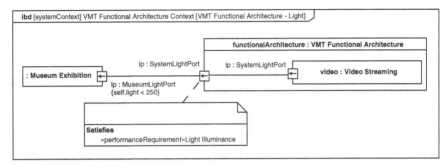

Figure 14.29. Nonfunctional requirements in the functional architecture.

architecture. The nonfunctional requirement states that the system must not produce illuminance equal or greater than 250 lux to protect light-sensitive museum artifacts. The light is an object flow in the appropriate use case activities and the functional architecture has ports with the light as a flow property. An additional constraint at the port ("self.light < 250") specifies that the illuminance of the light at the artifact is always < 250. The constraint is linked with the nonfunctional requirement by a satisfy relationship.

The UML Profile for Modeling Quality of Service and Fault Tolerance Characteristics and Mechanisms Specification defines a set of UML extensions to represent quality of service and fault-tolerance concepts [140].

There are also nonfunctional requirements that do not have a direct relationship to a functional requirement and therefore they are not covered by the functional architecture. For instance, the nonfunctional requirement that the color of the housing of the museum robots or that the mass of a single robot must not exceed 35 kg. Theoretically, you can treat a color as a function that transforms light or as a function for beauty. However, typically it is not valuable in practice. In consequence, there are some nonfunctional requirements that are not covered by the functional architecture. Be aware of them and consider them in the logical or product architecture. They could easily be detected by model queries that look for requirements without relationships to architecture artifacts.

14.11 COMPLETENESS OF THE FUNCTIONAL ARCHITECTURE

The first versions of the functional architecture will not be complete according to all required functional requirements of the system of interest. It is derived from use cases and typically use cases and the related use case activities do not cover all functions required to run the system.

First the use cases are not complete according to all functions that are related with system/actor interactions. Many projects only describe the main functions of a system with use cases. And some functions are not well suited to be refined with the use case concept.

Second there are more functions necessary to run the system. The infrastructure functions provide the fundamental for the actor-oriented functions. A common example is the power management of the system. Typically, those functions are only partly covered by use cases. For instance, "Deliver Power" is seldom described by a use case, although it is possible and could be valuable.

Third of course it is possible that not all functions requested by the stakeholders are covered by identified requirements. That is part of the requirements engineering discipline to complete the set of requirements and we do not elaborate that here further.

In the following, we will discuss how to increase the coverage of the system functions by the functional architecture. First we define three categories of functions (Figure 14.30):

A broad use case analysis covers most of the functions that have a direct effect on the interactions of the system with its environment. We call them as actor-oriented functions.

The second category are the infrastructure functions like functions of the power management.

The quality functions are the third category. They provide functionality to achieve quality that is requested by nonfunctional requirements. For instance, user authorization functions to assure data and operational safety. Quality functions could also be part of the actor-oriented or infrastructure functions.

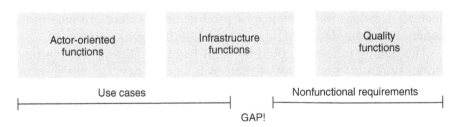

Figure 14.30. Functional coverage use cases and requirements.

There is no need to explicitly assign the functions to only one of the three categories. The categories are only a tool to support the identification of functions. Typically, there is a gap in the category infrastructure functions. The gap contains functions that are not covered by the requirements analysis, neither by a use case nor by a nonfunctional requirement. They could be spotted by the system architect or requirements engineer by analyzing the base architecture.

In the following, we describe methods to increase the coverage of the functions of each category in the functional architecture.

The actor-oriented functions could be completed by applying the following methods during the use case analysis:

1. Check the system context if it specifies a complete list of all human and nonhuman actors. Consider the active actors as well as the passive actors. The active actors initially contact the system. The passive actors are contacted by the system. Do not oversee environmental effects as a special actor category like temperature or humidity. They also interact with the system and could be related to system functions.

2. Check the system boundary and the ports of the system if they cover all data that goes into or out of the system. Also consider implicit data like heat or physical forces.

3. For each actor and port at the system boundary check if the in- and outgoing data is covered by a use case. If not you have probably found a new use case.

To complete the set of infrastructure functions: Analyze the infrastructure that is necessary to run your system. Keep care of the levels of the zigzag pattern and only consider infrastructure functions that are part of the base architecture (Section 7.2). A good example are functions for the power management. Typically, power management is part of a base architecture. Infrastructure functions that are based on architecture decisions in the logical or product architecture are part of the functional architecture on the next level of the zigzag pattern.

The quality functions could be identified by analyzing the nonfunctional requirements. Many of them will lead to new functions or use cases. Break them down until you have clearly separated the functional part from the real nonfunctional part.

Following these steps is not a guarantee that you cover all functions of your real system. No method could achieve that. But they could help to increase the coverage and to ask the right questions during your system analysis and architecture process.

14.12 FUNCTIONAL ARCHITECTURES AND THE ZIGZAG PATTERN

Functional architectures are on different levels according to the SYS-MOD zigzag pattern (see Section 7.1). If you decompose a function of a functional block, you need technical decisions at a certain level. Otherwise, further decomposition is not possible. The technical decisions lead to new requirements and functions that directly depend on the technical decisions.

The zigzag pattern describes different levels of requirements and architectures. An architecture decision leads to new requirements and the requirements to new architecture decisions and so on. See Section 7.1 for a detailed description of the zigzag pattern.

Figure 14.31 shows the requirement collision avoidance of the VMT. The requirement "Robot Collision Avoidance" is refined by some use cases for instance "Steer the robot" as shown in Figure 14.31. Some of the activities of the use case are grouped in a functional group "Robot Control" that leads to a functional block of same name. The functional block is allocated beside others to the "Anti-Collision System" (ACS). The ACS is modeled as a variation point to specify different variants of the ACS. The variant of a camera-based collision system leads to new requirements and opens the next level of the zigzag pattern (Figure 14.32). The camera-based ACS is a specialization of the variation point ACS. It has its own functional architecture with functional blocks "Imaging" and "Communication." The functional blocks have relationships to the functional blocks on the upper zigzag level. The relationship is depicted in Figure 14.33.

You can treat the ACS like a system with its own requirements, use cases, and architectures including a functional architecture.

228

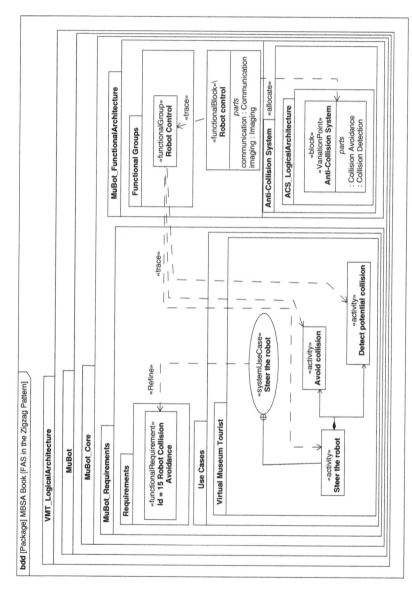

Figure 14.31. Example FAS in the zigzag pattern.

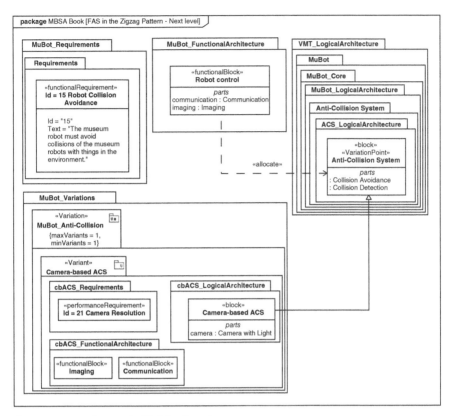

Figure 14.32. Example FAS in the zigzag pattern.

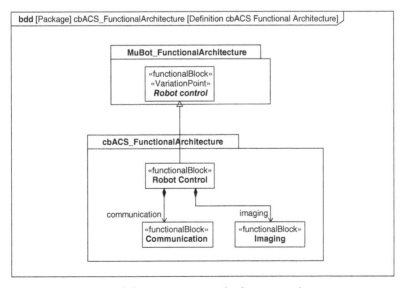

Figure 14.33. Functional decomposition of robot control.

Chapter 15

Product Lines & Variants

Many systems exist in different configurations: A product line, a customized product, or different designs for trade studies. Typically, a single variant of a system affects only a few parts of the system. It is a slight derivation from the initial system. However, it is not possible to quantify the number or level of detail that varies to be still a variant of a system and not a complete new system.

A car as well as an aircraft could be a variant of a transportation system. However, in most cases, it makes no sense in practice to handle a car and an aircraft as variants of the same system and to manage all the appropriate relationships in a single system model. The common parts of a transportation system are too abstract. Unfortunately, you can't measure abstraction and we cannot give an objective metric. You must decide if the abstraction levels of the common parts and the abstraction level of the variant parts are close enough to be valuable to be part of the same model. The benefit must be larger than the effort to manage a complex model.

The description of variants is a sophisticated task. It is already challenging to create a good description of a single system. Each variation adds another dimension to a multidimensional system model. For instance, the engine could be a variation of a car system with three possible variants: diesel, electric, or hybrid engine. Next variation could be the chassis: small, deluxe, or cabriolet. Now you can combine the variants, for example, a car with diesel engine and a small chassis or

Model-Based System Architecture, First Edition.
Tim Weilkiens, Jesko G. Lamm, Stephan Roth, and Markus Walker.
© 2016 John Wiley & Sons, Inc. Published 2016 by John Wiley & Sons, Inc.

a car with a hybrid engine and a deluxe chassis, and so on. Any additional variation increases the dimension and the number of potential combinations.

In the following, we will first give some definitions and then describe a concept for variant modeling with SysML. Parts of this chapter are based on the book Variant Modeling with SysML by our author Tim Weilkiens [148]. We omit the citation of the book at each statement that is taken unchanged or updated from that book.

15.1   DEFINITIONS VARIANT MODELING

We give some brief definitions of the terms how we use them in the context of variant modeling. Our concept of variant modeling is described in the SYSMOD methodology [145, 147] and later refined in the book "Variant Modeling with SysML" [148]. The terms and concepts are conformed with common variant concepts presented in publications about variant modeling, for instance, the orthogonal variability model (OVM) [118].

Figure 15.1 shows the domain model of the variant modeling method. We differentiate between core and variant elements. A *core element* is used in all system combinations and is independent of any variant elements. A *variant element* only occurs in some configurations and is part of a variant.

A *variation point* marks a core element of the system as a docking point for a variant element, for example, the engine of a car. It is a core element since any of our cars have an engine. And it is a variation point. We vary that part of the system to provide different kinds of engines.

The reason for a variant is called *variation*. A variation contains a set of variants that have a common discriminator. In our car example, the engine type is a variation. It is the discriminator that distinguishes into the variants.

A *variant* is a complete set of variant elements that varies the system according to the variation. A variant is also known as a feature of the system. The diesel, electric, and hybrid engines are variants of the variation engine kind.

A *variant configuration* is a valid set of variants and the core, for instance, a car with a hybrid engine and a deluxe chassis. A variant configuration is also a special variant and part of a variation. In our example, the variant configuration "hybrid engine with deluxe chassis" is part of the variation "eco editions."

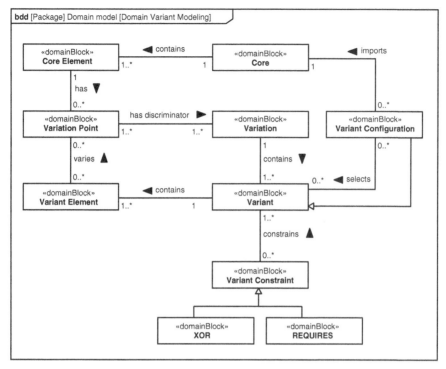

Figure 15.1. Domain model for variant modeling.

A *variant constraint* specifies rules for a valid set of variants. Two common variant constraints are predefined. *XOR* is used to exclude a variant if a specific variant is selected. *REQUIRES* specifies that other variants are required if a specific variant is selected.

A variant could again include variations (Figure 15.2). The structure of these variations is the same as for the top-level variations. This recursive structure makes the variant modeling concept scalable for any size of a system.

15.2   VARIANT MODELING WITH SysML

SysML does not provide explicit built-in language constructs to model variants. However, SysML is useful to model variants and you can use the profile mechanism of SysML to extend the language with a concept for variant modeling. SYSMOD defines a profile for SysML that covers variant modeling as presented in the previous section [145]. The profile provides stereotypes to model the concepts of variants, variations,

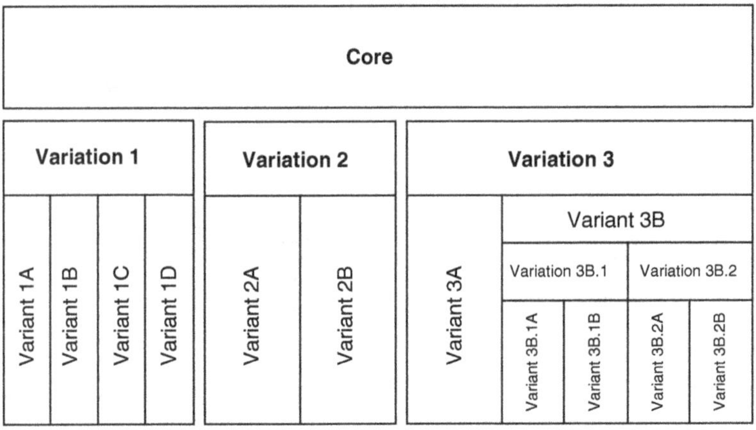

Core											
Variation 1				**Variation 2**		**Variation 3**					
							Variant 3B				
							Variation 3B.1		Variation 3B.2		
Variant 1A	Variant 1B	Variant 1C	Variant 1D	Variant 2A	Variant 2B	Variant 3A	Variant 3B.1A	Variant 3B.1B	Variant 3B.2A	Variant 3B.2B	

Figure 15.2. Core and variants.

variation points, variant elements, variant configurations, and variant constraints.

Our VMT example system has different variations, for instance the anti-collision and the visual system of the museum robots. Figure 15.3 shows the top-level package structure of our model with the packages for the variant modeling. On the first level, we have three packages:

- The configurations package ("MuBot_Configurations") contains the variant configurations, that is, valid sets of core and variant elements combined to a system or system assembly. Since a variant configuration is also a special variant we have variation packages on the first level and the variant configurations as variants on the next level. The structure of variation and variant packages is described later.
- The core package ("MuBot_Core") contains all core elements. The structure of the subpackages conforms to the system model structure presented in Section 7.9.
- The variations package ("MuBot_Variations") contains all variations with their variants. In Figure 15.3, you see three variations: "MuBot_Anti-Collision," "MuBot_Chassis," and "MuBot_Visual System."

The top-level variations package contains the variations. A variation is a package with stereotype «variation». Each variation package contains the variants according to the variation discriminator.

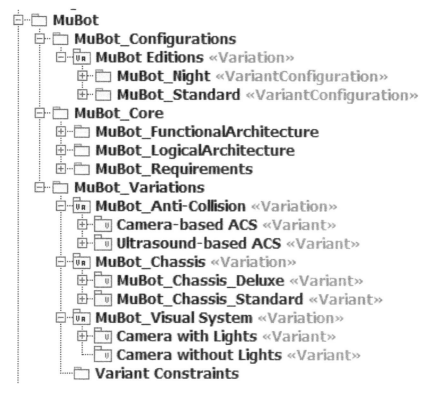

Figure 15.3. Top-level package structure.

The variation package has two additional properties ("minVariants"and "maxVariants") that constrain the number of variants of the variation that could be selected for a single variant configuration. Figure 15.4 depicts the package structure in a SysML package diagram with the "minVariant" and "maxVariant" specifications shown in the package symbol of the variations. For the anti-collision system (ACS) we allow only one variant per variant configuration ("maxVariants=1"). If for instance if "maxVariants" would be 2, it is allowed to implement two different ACS into a single museum robot. The ACS is mandatory. That is specified by "minVariants=1," that is, at least 1 variant must be selected for a valid variant configuration.

A variant is a package with stereotype «variant». The variant package is the root for all variant elements. They are organized like the recursive package structure for system models (Section 7.9). A variant could be handled like a system or subsystem with a context, requirements, architectures as well as again configurations, variations, and variants.

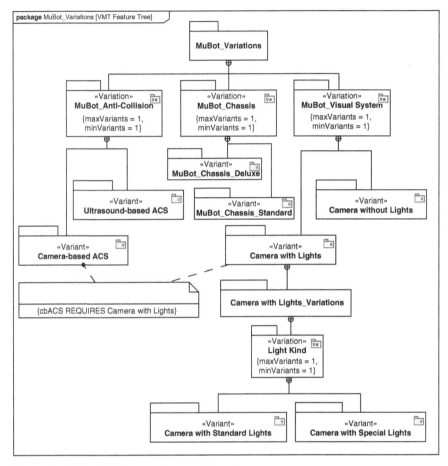

Figure 15.4. SysML feature tree with variants.

Maybe you already know the feature trees to describe variations. For instance, from the feature-oriented domain analysis (FODA) [70]. See more details about FODA and an example of a feature tree in Section 15.3. You can also use SysML with the variant profile to create feature trees as you can see in Figure 15.4. The notation is different from FODA, but the semantic is conformed to FODA trees. Feature trees specify rules between variants. For instance variants that excludes each other or variants that require other variants to be part of the system.

XOR and *REQUIRES* are special variant constraints to model rules between variants of the same or different variations. Besides the ACS, the museum robot has variants for the visual system and the chassis. A camera system without additional lights, a camera system with standard lights, and a camera system with special lights to protect sensitive museum artifacts. A museum robot could have only one of these visual systems. The camera-based ACS requires a camera with lights to assure the detection of obstacles even in dark areas of the museum. Figure 15.4 depicts the appropriate variant constraint. The *REQUIRES* constraint assures that a camera-based ACS is always combined with a visual system with lights. Since we have two visual system variants with lights, we have introduced another variation "Light Kind." Alternatively, we could introduce a new variant constraint (A REQUIRES B) OR (A REQUIRES C). The additional variant level is more flexible, because it is not necessary to change the variant constraint if we define another visual system variant with lights and must combine A, B, C, and D in a single constraint. Formally, the variant constraints are stereotypes («xor», «requires») of the general SysML constraint element.

The chassis variants in Figure 15.4 depend on business requirements. The deluxe chassis is more durable and necessary for intensive use of the system. The standard chassis is for normal use. The relationship between the variants and the business requirements is in the model, but not depicted in Figure 15.4.

There are relationships between variant and core elements. The variant elements always depend on the core elements and not vice versa. Typically, the relationship is a generalization, that is, a core element is a generalization of a variant element, respectively, the variant element is a specialization of the core element.

Figure 15.5 depicts the variant element "Camera-based ACS" as a specialization of the core element "Anti-Collision System." And the variant element "Camera with Light" as a specialization of the core element "Camera." The variant elements are part of different variants. The appropriate variant "Camera-based ACS" requires the variant "Camera with Light" which is reflected not only by the *REQUIRES* constraint, but also by the composition relationship in Figure 15.5.

Figure 15.6 depicts the variant configuration "Logical MuBot Night." The main task of a variant configuration is to bundle variants and the core to a valid assembly. The block "Logical MuBot Night" specializes the core element "Museum Robot" from the logical architecture and links the root elements of the variants camera-based ACS and camera with standard lights. In addition, a variant configuration could define its own structure and behavior. Typically, it is the glue logic to assemble the core and selected variant elements.

The variant configuration is a stereotype «variantConfiguration» specializing the stereotype «variant» and be applied to a SysML

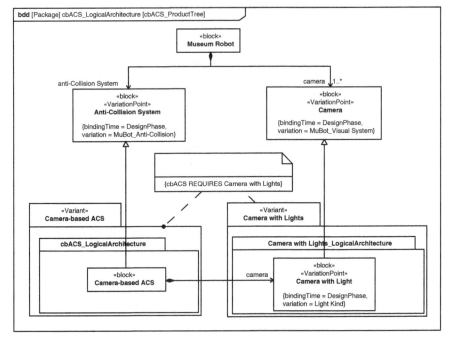

Figure 15.5. Relationship between variant and core elements.

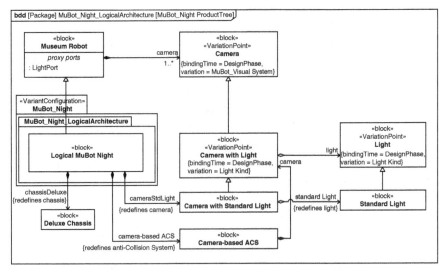

Figure 15.6. Variant configuration.

package. A variant configuration could bundle a system, a subsystem, or another system unit. Variant configurations are stored as special variant packages below a variation package in the top-level configuration package. Each variant configuration has a package structure like the system model (Section 7.9).

Although all these stereotypes are simple and powerful it's a challenge to handle the complexity of the model. Even a system model without variants could already be a challenge. With variants it is a multidimensional configuration space. Special views, reports, and model transformations are necessary to manage the complexity.

15.3   OTHER VARIANT MODELING TECHNIQUES

In this chapter, we will look a little bit out of the box and briefly describe some other variant modeling techniques.

A common technique to model variability is the FODA by Kang et al. [70]. The variability is modeled from the perspective of the stakeholders. It shows the features of the system, their variability, and constraints between the variants. Figure 15.7 shows a FODA feature tree for the VMT. It depicts the three features ACS, engine, and human interaction package of the museum robot.

Later Kang extended FODA to the feature-oriented reuse method (FORM) that in addition to the requirements and architecture perspective incorporates the marketing perspective [71].

There are two different basic approaches for variant modeling to integrate the variability information into the system model:

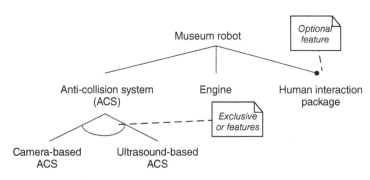

Figure 15.7. Example FODA tree.

1. Create a separate orthogonal first-class model that specifies the variability.
2. Integrate the variability into the model of the system under development.

Examples for the second approach could be found in Ref. [71] and for the first approach in Ref. [118]. The approach presented in Sections 15.1

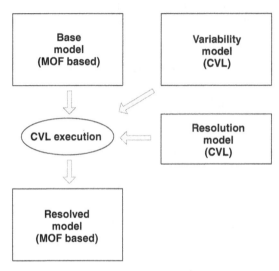

Figure 15.8. CVL concept [102].

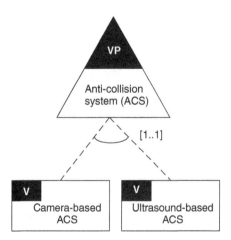

Figure 15.9. Example OVM.

and 15.2 is another example for the first approach. The variability information is separated from the development model and could even be stored in a separate repository.

The Common Variability Language (CVL) [102] is an upcoming standard from the Object Management Group (OMG). It is a model language only to model the variability aspect and not the complete system. The CVL model is applied to a model based on the Meta Object Facility (MOF). MOF is another standard from the OMG [104] and languages like UML and SysML are built on MOF. The application of a CVL model to a MOF-based model leads to a product model (similar to the variant configuration described earlier) (Figure 15.8). CVL is not yet officially released by the time of writing this book, and to our knowledge, there are no commercial modeling tools for CVL available.

Another variant modeling technique is the OVM [118]. Figure 15.9 shows an extract of our FODA tree in Figure 15.7 in OVM language.

Chapter 16

Architecture Frameworks

The term architecture framework is frequently misunderstood. It is often assumed that an architecture framework provides a kind of abstract or common template for a systems architecture. Something like a "skeletal architecture" of a certain kind of system, whose gaps just must be filled to get a full architecture for a system. For example, one could guess that the general architecture for an aircraft, like body, wings, propulsion subsystem, landing gears, power, and energy subsystem, and so on is abstractly defined in a kind of "Architecture Framework for Aircrafts", and the systems engineer could use that framework as a starting point for an aircraft systems architecture development.

But that's not correct.

Of course, there are proven base architectures, sometimes also called reference architectures, for miscellaneous kinds of systems in the systems engineering domain, but architecture frameworks didn't define them!

But what are architecture frameworks instead? The international standard ISO/IEC/IEEE 42010:2011 *Systems and software engineering — Architecture description* [64] defines architecture framework as conventions, principles and practice to describe an architecture. An architecture framework according to the standard is for a specific domain of application or community of stakeholders.

Model-Based System Architecture, First Edition.
Tim Weilkiens, Jesko G. Lamm, Stephan Roth, and Markus Walker.
© 2016 John Wiley & Sons, Inc. Published 2016 by John Wiley & Sons, Inc.

As a first notice about this definition is that it is about the description of architectures. So that these architecture descriptions comply with certain standards, architecture frameworks provide a set of conventions, principles, and practices, which can be used by systems engineers to create such descriptions. Creating and describing of a systems architecture from scratch can be a daunting task; therefore, architecture frameworks can be helpful and could provide a good starting point. They should simplify the process and guide an architect through all areas of architecture development.

Another noteworthy detail is that this definition makes no specific statement about the kind of system whose architecture shall be described. It could be a technical system architecture, a software architecture, an enterprise architecture (EA), or a system of systems architecture, just to mention a few possibilities. The definition speaks quite general about "established within a specific domain of application and/or community of stakeholders". What is considered to be "established" in certain domains may be very different.

In practice, architecture frameworks are very well known in the world of enterprise architecture development, and system of systems engineering (SoSE). There are a number of different frameworks for SoSE, especially in the defense domain. Before we discuss those frameworks in detail, we first have to discuss the term enterprise architecture; afterward, I would like to take you on a short excursion to the noteworthy characteristics of a system of systems.

16.1 ENTERPRISE ARCHITECTURES

If architecture frameworks are frequently used in the world of Enterprise Architecture development, we first have to define the term enterprise architecture (short: EA). In fact, there is no generally accepted definition of the term enterprise architecture. Various organizations (public and private) promote their understanding of the term. For example, the *Federation of Enterprise Architecture Professional Organizations* (FEAPO) defines it the following way:

> Enterprise Architecture is a well-defined practice for conducting enterprise analysis, design, planning, and implementation, using a holistic approach at all times, for the successful development and execution of strategy. Enterprise Architecture applies architecture principles and practices to guide organizations through the business, information, process, and technology changes necessary to execute their strategies.

These practices utilize the various aspects of an enterprise to identify, motivate, and achieve these changes [37].

But what is meant with an enterprise? Broadly speaking, the term enterprise covers various types of organizations. The size of the organization, its industry, its ownership model (private or public), or its geographical distribution doesn't matter. Enterprise can mean a small to mid-size company, a multinational corporation, or a huge joint undertaking with a common goal shared by several global players. An enterprise can be a peacemaking and peacekeeping deployment of North Atlantic Treaty Organization (NATO) forces, or an outer space research program of the NASA. And of course, the management, operation, and further development of the museum from our accompanying case study in this book can be considered as an enterprise.

Figure 16.1 depicts the way how a technical systems architecture is derived from the stakeholder's business goals and visions. It is not sufficient that individual projects or the product development just fulfill immediate stakeholder needs. Instead, a long-term view on all processes, systems, and technologies of an enterprise is required,

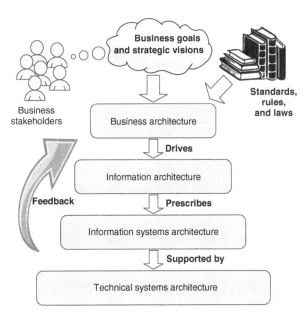

Figure 16.1. The interrelationship between business goals, information architecture and technology environments.

so that individual projects or the product development can provide general-purpose capabilities. This is important for investment decision-making, work prioritization, and resource allocation. In our world of ever-increasing complexity, ever-changing business environments, and globalization, it's not only of huge importance that an enterprise is able to appropriately respond to disruptive forces. Furthermore, every enterprise has its visions and goals that adapt a predicted future. This requires a holistic view on the enterprise, and a departure from an isolated treatment of individual units, for example, business segments, single departments, or solely the IT infrastructure. Enterprise architecture development is a business-centric approach, and not a technology-centric one.

Enterprise architectures (EA), and thus the discipline of Enterprise Architecture development, has the following major goals:

1. The architecture of an enterprise, its units, policies, processes, strategies, and technological infrastructure (e.g., IT systems), supports all stakeholders in achieving short- and long-term business objectives and visions.
2. EA fosters an alignment of the technological systems developed by, respectively used by an enterprise with its business goals and strategic direction.
3. EA helps an enterprise to learn, grow, innovate, and respond to market demands and changing basic conditions.
4. EA fosters and maintains the learning capabilities of enterprises so that they may be sustainable.

Of course, it strongly depends on the domain what is meant by the notions *business objective* and *business vision*. The business objectives of an enterprise in the defense sector, or in the space domain, are usually different than in the commercial business environment (e.g., insurance companies, world of finance, or trading). But on an abstract level it's all the same: all decisions, activities, developments, and procurements that are done by an enterprise must support these goals, and the discipline of enterprise architecture development is the key.

16.2 SYSTEM OF SYSTEMS (SoS)

Virtually, no system is developed on a greenfield, neither it is used independently from other systems. Every system has a context, that is, on the one hand it interacts with humans, and on the other hand with systems in its environment. And systems are sometimes used in conjunction with many other technical and socio-technical systems as part of a larger, much more complex system. Under certain circumstances, such kind of compositions of multiple, dispersed, and independent systems is called a system of systems (SoS).

A system of systems is different from a single monolithic system, and from other kinds of system groupings (e.g., a family of systems), in the following manner:

1. *Loose coupling of elements:* In general, the term coupling refers to the degree that constituents depend on each other. One distinguishing feature, which lets you know that you are confronted with a system of systems instead of a single monolithic system, is the remarkable loose coupling of its constituents, that is, of its component systems. The component systems do not interchange forces, fluids, or large amounts of energy with each other. The coupling between them usually takes place via data exchange through IT interfaces. This is usually accompanied by geographical distribution of the component systems.

2. *Emergent behavior:* The system of systems behavior arise from the cumulative actions and interactions of its constituents. A system of systems behavior is more than just the sum of the capabilities of its component systems. To create emergent behavior, all component systems must interact with each other. A prerequisite is that each system can exchange messages with other systems. Since emergence is a key characteristic of SoS, the term is deepened in the following subsection.

3. *Operational independence:* If a system of systems is disassembled into its constituents (component systems), these systems are able to operate independently outside of a system of systems compound, or they can join another system of systems.

4. *Independent management of elements:* Most of the component systems that are building a system of systems are independently developed, manufactured, purchased, and administrated. Their life cycle is independent from the life cycle of the whole system of systems and from that of the other constituent systems.

5. *Evolutionary development:* A complete fully developed system of systems does typically not exist. Its development and existence is evolutionary, that is, over the entire life cycle component systems

can be added, removed, modified, or exchanged, depending on new or changed requirements, new capabilities that are required, altered basic conditions, or changed objectives.

Although not every system of systems have all of the aforementioned properties, these criterions are typical characteristics of systems of systems. The most outstanding characteristic of a SoS are its emergent properties. In this respect, a SoS differs most significantly from the concept of a family of systems. With a family of systems, no capabilities beyond the individual functionalities of the joint component systems are provided by the merger.

16.2.1 Emergence

Emergence was first defined properly by the English philosopher George Henry Lewes (1817–1878) in his 1875 published work *Problems of Life and Mind* [87]:

> Every resultant is either a sum or a difference of the co-operant forces; their sum, when their directions are the same—their difference, when their directions are contrary. Further, every resultant is clearly traceable in its components, because these are homogeneous and commensurable. It is otherwise with emergents, when, instead of adding measurable motion to measurable motion, or things of one kind to other individuals of their kind, there is a cooperation of things of unlike kinds. The emergent is unlike its components insofar as these are incommensurable, and it cannot be reduced to their sum or their difference.

The term emergence refers to the occurrence of new properties from the interaction of the elements in a system, whereby these emergent properties cannot be derived directly from individual parts of the system. Even with perfect knowledge about the components, some properties of the entire system cannot be predicted.

Emergent properties can be very well observed in nature. Good examples are the shape and behavior of a flock of birds or school of fish. From a human's perspective, a flock of birds or a school of fish behaves like a single, large organism. But their individuals only interact with a small amount of other group members in their closest neighborhood. The single bird, or fish, know nothing about the shape and position of the whole swarm. Through emergence collective properties arise, that cannot be localized in the constituents of the swarm.

Another good example for emergent behavior is a colony of ants. The individual ants do not know anything about coordinated foraging. A single ant wanders in search of food around randomly, as biologists have discovered. But suddenly, at a certain point, the movement patterns of the ants change from chaos to order. It seems that the whole colony has a common strategy for coordinated foraging, and the collective behaves as a highly efficient complex network.

Even in the man-made civilization, there are numerous examples of emergence. For instance, the complex business world of today knows, and fears emergent properties. There, the term is often associated with unintended effects that may arise from different and superimposed management decisions in large enterprises. Suddenly something happens that no one has foreseen, or more precisely: could not foresee. And these effects are increasingly becoming a problem for traditionally organized, hierarchical organizations, because more and more frequently these effects bring rather damage to the company. The same applies to the globally interconnected, complex financial world.

Even in Systems Engineering, such arising emergent properties can be both, intended and unintended. Quite often, especially unwanted emergent behavior arise after the integration of a system. All elements that compose the system work correctly for themselves. But if these elements are assembled to a system, this system may show undesirable behavior.

A system of systems depends on emergent behaviors to achieve its goals. It is explicitly desired that a SoS evolves certain emergent properties. A special challenge for the System of Systems Engineer is to create an architecture in such a way, that these properties will emerge purposefully. That could be a difficult task, because emergent composition is often poorly understood. Just imagine that you should build a highly reliable system of systems from unreliable component systems.

16.3   AN OVERVIEW OF ARCHITECTURE FRAMEWORKS

As already mentioned above, there are a variety of different architecture frameworks in existence. They differ mainly with respect to their field of application. Some have been created in the military environment, some other in the domain of enterprise architecture IT development, and some are especially designed for complex SoSE in the space domain.

Several architecture frameworks are defining a set of perspectives and views (The concept of views is discussed in detail in Section 6.2.2) that should be created by the SoS engineer. For this purpose, these architecture frameworks specify a UML-based meta model which defines appropriate model elements and relationships. These meta models are usually available as a stereotype profile for importing into modeling tools. Some frameworks also specify an enterprise architecture development process respectively methodology.

The following subchapters provide an overview and lists the most important properties of a couple of best-known architecture frameworks.

16.3.1 Zachman Framework™

Brief Description	
First year of publication	1987
Developed, respectively published by	John A. Zachman, Zachman International® Inc.
Primary field of application	IT enterprise architecture
Development process or methodology	no
Meta model (Ontology)	no

The *Zachman Framework for Enterprise Architecture and Information Systems Architecture* [151], better known under its short name *The Zachman Framework^{TM}*, was originated by John A. Zachman, a US-American business and IT specialist, in 1987. It is considered as one of the most important frameworks and influenced the contemporary understanding of enterprise architectures significantly. Many subsequent developed enterprise architecture frameworks have had the Zachman Framework as a foundation.

The framework defines prestructured views and layers to represent an information technology (IT) enterprise. Unlike similar frameworks that often contain process models, the Zachman Framework does not prescribe any process or methodology. It focuses on the roles involved and assigns them to objects that shall be viewed from different perspectives. The Zachman Framework thereby provides a comprehensive tool to consider all relevant aspects, from all perspectives, while designing and developing an enterprise IT architecture.

Although this framework plays virtually no role in Systems Engineering, it is regarded as one of the pioneer works in the area of architecture frameworks.

16.3.2 The Open Group Architecture Framework (TOGAF®)

Brief Description	
First year of publication	1995
Developed, respectively published by	The Open Group
Primary field of application	Enterprise architecture
Development process or methodology	Architecture Development Method (ADM)
Meta model (Ontology)	No

First published in 1995, *The Open Group Architecture Framework* (TOGAF®)[136] was based on the *US Department of Defense Technical Architecture Framework for Information Management* (TAFIM; a framework that is not discussed here). It has been developed by The Open Group, a vendor and technology-neutral industry consortium. The framework provides an approach for designing, planning, implementing, and governing an Enterprise IT architecture. The latest version is TOGAF 9.1, published on December 1, 2011.

Unlike many other frameworks, the TOGAF standard provides a detailed process model. The Architecture Development Method (ADM, see Figure 16.2) is an iterative process over all phases of enterprise architecture development, adaptable to specific needs.

Tailoring the ADM to support specific needs is, among other preparations, something that should take place in the Preliminary phase. In phase A, stakeholders are identified, and the scope, constraints, and expectations are defined. Furthermore, a vision of the project is developed. The core phases of enterprise architecture development are the phases B, C, and D. In these three steps, the Business Architecture, the Information Systems Architecture, and the Technology Architecture are developed. Another step (phase F) is planning the migration, that is, how to move from the Status Quo (the current architecture of the enterprise) to the target architecture. Phase G ensures that the implementation project conforms to the planned architecture. Every step has a bidirectional relationship with the Requirements Management in the

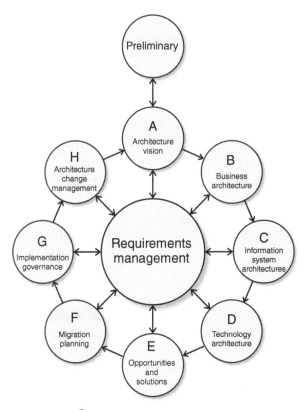

Figure 16.2. The TOGAF® Architecture Development Method (ADM).

center of the ADM diagram. This is to emphasize the importance of requirements in each step, and that new or changed requirements can be discovered every time due to new insights.

In contrast, the TOGAF standard does not define or prescribe form and look of its process outcomes, that is, the deliverables that must be built in every phase during development. For instance, the form how the Technical Architecture in step D has to be documented, respectively visualized is not defined. It could be a simple document containing a textual description, a textual and graphical description in a Wiki, or a model of the technical architecture in a modeling tool. For this reason, TOGAF is often combined with other view-oriented frameworks that have a UML-based meta model, which in turn does not have a full-fledged process or methodology.

Although the TOGAF standard is preferably intended for the development of enterprise IT architectures, the framework could also be used in the Systems Engineering domain. For instance, the *Norwegian*

Armed Forces chose TOGAF's ADM as their architecture development methodology, and the NATO Architecture Framework (NAF, see Section 16.3.5) for meta model and content organization (Views) [89].

16.3.3 Federal Enterprise Architecture Framework (FEAF)

Brief Description	
First year of publication	1999
Developed, respectively published by	U.S. Chief Information Officers (CIO)
Primary field of application	Enterprise architecture of Federal Governments
Development process or methodology	Collaborative Planning Methodology (CPM)
Meta model (Ontology)	Consolidated Reference Model (CRM)

In the United States, the Chief Information Officers (CIO) serve as a central resource for information on federal information technology (IT). In 1996, the *Clinger–Cohen Act* (CCA), formerly known as the *Information Technology Management Reform Act*, was enacted by the U.S. Congress to reform and improve the way federal agencies acquire and manage IT resources. Thereupon in September 1999, the CIO published an architecture framework that should support federal authorities and agencies in the development of Enterprise IT Architectures: the *federal Enterprise Architecture Framework* (FEAF) version 1.1. In January 2013, a revised version 2.0 of FEAF was published [135].

As its name suggests, FEAF is primarily intended for the Enterprise Architecture development in federal agencies. The framework should standardize the development and use of architectures within and between those federal agencies. FEAF provides both, a structure (Consolidated Reference Model (CRM)) and a methodology (Collaborative Planning Methodology (CPM)). The CRM consists of six reference models and provide standardized categorization for strategic, business, and technology models. The CPM is a full planning and implementation life cycle for federal enterprise architectures. It consists of two main phases: Organize and Plan, and Implement and Measure. These main phases are further divided into smaller steps. Although the CPM looks strictly sequential and reminds of a waterfall-like approach, it is emphasized that there are frequent iterations within and between the phases.

It is not known whether FEAF was also used for any SoSE architecture development purpose. The development of FEAF demonstrates, however, that very special requirements and needs in a certain domain may result in a highly specialized architecture framework.

16.3.4 Department of Defense Architecture Framework (DoDAF)

Brief Description	
First year of publication	2003
Developed, respectively published by	U.S. Department of Defense
Primary field of application	Military operations and System of Systems
Development process or methodology	Six-step Architecture Development Process
Meta model (Ontology)	DoDAF Meta Model (DM2), based on UML

The predecessor of the *U.S. Department of Defense Architecture Framework* (DoDAF) was the *C4ISR Architecture Framework* (C4ISR AF, a framework that is not discussed here) version 2.0. C4ISR is a military acronym for command and control, communications, computers, intelligence, surveillance, and reconnaissance. The term refers mainly to the interconnection of all management, information, and monitoring systems in order to create a more accurate picture of the overall situation, and thus to improve decision-making and leadership skills.

DoDAF 1.0 replaced its predecessor in August 2003. There followed an evolutionary development of the framework, and ended in DoDAF 2.0, which was released in May 2009. One of the major changes between DoDAF version 1.0 and 2.0 was the transition from a product-centric process to a data-centric process. The current version is DoDAF 2.02 [141].

DoDAF is a view-oriented architecture framework, that is, it provides an organized, meta-model-based visualization infrastructure for specific stakeholders concerns. Therefore, the framework consists of eight main viewpoints, whereby the meaning of the term viewpoint is consistent to its meaning in SysML. A viewpoint defines specifies rules for constructing a view (perspective) on the system under development including a set of concerned stakeholders and the purpose to address their concerns.

Each one of the eight viewpoints consists of a coherent set of views representing the architecture of the enterprise from the perspective of the viewpoint. For example, DoDAF's Operational Viewpoint (OV) consists of nine different views (OV-1 … OV-6c), all together describing operational scenarios, activities, and requirements that are necessary to develop those capabilities that must emerge through the planned System of Systems. The whole DoDAF 2.0 framework consists of 52 views.

All DoDAF views must follow a UML-based meta model (DoDAF Meta Model; short: DM2) which defines specific types, semantics, relationships, rules, and constraints about how to build a DoDAF-conformant system of systems enterprise architecture model. For many well-known UML/SysML modeling tools on the market, this meta model is available as an UML 2 stereotype profile, respectively a plug-in. The *Unified Profile for DoDAF/MODAF* (UPDM, see Section 16.4) contains all stereotypes required for DoDAF modeling.

In contrast to its very detailed meta model, DoDAF contains only a high-level six-step phase model for dealing with the development process of an enterprise architecture. If a more detailed methodology is required, a mapping of DoDAF 2 views to TOGAF® ADM deliverables and activities might be a feasible solution.

16.3.5   NATO Architecture Framework (NAF)

Brief Description	
First year of publication	2004
Developed, respectively published by	North Atlantic Treaty Organization (NATO)
Primary field of application	Military operations and complex System of Systems
Development process or methodology	NC3A Architecture Engineering Methodology (AEM)
Meta model (Ontology)	NAF Meta Model (NMM), identical to MODAF M3

First published in September 2004, the NAF is an enterprise architecture framework used by the NATO to define the operational context, the system architecture, and the supporting standards and documents that are necessary to describe an enterprise. The framework is broadly accepted and used in the defense domain. The main objective of NAF is the successful exchange of architecture-data between all stakeholders in the context of joint missions of the intergovernmental military alliance, NATO.

The first version of NAF was developed on the base of DoDAF (see Section 16.3.4). As DoDAF it is a view-oriented framework, whereby the views are called "view templates." Since the current version NAF 3.1 is still based on DoDAF for backward compatibility, but especially the meta model has become influenced by MODAF (see Section 16.3.6). Version 3.1 of the NATO Meta Model is identical to version 1.2.003 of the MODAF Meta Model (M3).

The whole framework is described in a set of documents, consisting of an Executive Summary, NAF Chapters 1–7, and three annexes (A … C). The most interesting chapters for modelers are Chapter 4, which describes all main views and subviews, and Chapter 5 that describes the meta model.

Even NAF does not prescribe any process or methodology. Annex B presents several process models and methodologies that can be used with NAF, for example, TOGAF ADM, DoDAF's six-step model, or NC3A's Architecture Engineering Methodology (AEM). The AEM is fully described.

NAF is seen as a key enabler for effective communications about the results of an architecture development process. The ability to effectively federate capabilities in coalition operations is crucial for the success of joint operations of NATO forces. An architecture development using NAF should therefore ensure that various component systems of different armed forces are working optimally together, and let emerge the capabilities of the system of systems compound as expected by the stakeholders.

Meanwhile NAF is also used, often in a customized form, for several nonmilitary projects and enterprises. One example is the *Single European Sky ATM Research Program* (SESAR). SESAR is a joint undertaking initiated by the *European Commission* and the *European Organization for the Safety of Air Navigation* (EUROCONTROL). Its goal is to completely overhaul the European airspace and its air traffic management. SESAR uses a framework during its development phase that is based on NAF and MoDAF but was customized to adapt aspects of them to meet Air Traffic Management (ATM) needs.

A version 4.0 of NAF is currently under development. This new version will have a new meta model called MODEM (*MODAF Ontological Data Exchange Model*) that was originally developed by the Swedish Armed Forces and the British Ministry of Defense. MODEM is an IDEAS-model, that is, it is based on the multinational *International Defense Enterprise Architecture Specification* (IDEAS) ontology that is defined as a formal ontology to facilitate interoperability of Enterprise Architecture models. IDEAS is defined as a UML stereotype profile and can be considered as a kind of meta-meta-model, that is, it is a basic framework for expressing concepts that are frequently used in enterprise architectures, and thus must often be represented as meta-classes in meta models of architecture frameworks like NAF. More information on NAF version 4.0 and its stage of development are available on its website [100].

16.3.6 Ministry of Defense Architecture Framework (MODAF)

Brief Description	
First year of publication	2005
Developed, respectively published by	British Ministry of Defense (MOD)
Primary field of application	Military operations and System of Systems
Development process or methodology	MODAF Architecting Process
Meta model (Ontology)	MODAF Meta Model (M3), based on UML

The *British Ministry of Defense Architecture Framework* (MODAF) [46] is a view-oriented architecture framework that is based on the DoDAF version 1 baseline. MODAF version 1.0 was released in August 2005. The last version 1.2.004 was released in May 2010.

As well as with DoDAF and NAF, each viewpoint of MoDAF offers a different perspective on the system of systems project to support different stakeholder concerns and interests. MODAF 1.2.004 consists of seven different viewpoints: All Views viewpoint, Strategic View viewpoint, Operational View viewpoint, System View viewpoint, Technical Standards View viewpoint, Acquisition View viewpoint, and Service-Oriented View viewpoint. All these viewpoints contain a different number of views, in total there are 47 views.

The MODAF Meta Model (M3) defines a UML2-conformant stereotype profile that specifies the structure of the architectural information that is presented in the MODAF views. Even the standardized *Unified Profile for DoDAF/MODAF* (UPDM, see Section 16.4) supports the modeling of MODAF-conformant enterprise architectures.

MODAF does not prescribe an official architecting process or methodology. However, there is a paper on the MODAF website available that describes a six-step approach named "The MODAF Architecting Process", which is just an example of one way to take.

16.3.7 TRAK

Brief Description	
First year of publication	2010
Developed, respectively published by	London Underground Limited/UK Department for Transport
Primary field of application	Each kind of complex system or System of Systems
Development process or methodology	No
Meta model (Ontology)	TRAK Meta Model, based on MODAF Meta Model (M3)

TRAK [116] is an interesting example for an architecture framework that was released under an Open Source License. Its logical definition is released under the GNU Free Documentation License (GFDL). Implementations of TRAK, for example, plug-ins or stereotype profiles for modeling tools, are licensed under a GNU Public License (GPL).

Furthermore, TRAK is a framework that is especially targeted to Systems Engineering. Its development started in 2009 under the working title *The Rail Architecture Framework* (acronym: TRAK) because it was thought that the result may be specific to the railway transportation domain. The reason for that was its development based on the then current views of architectural description within London

Underground, a public rapid transit system in the capital city of the United Kingdom, which is also known as "The Tube." However, it quickly became apparent that the finally developed framework will be domain-agnostic, that is, it contains no rail-specific views or meta model elements. This is one of the reasons why today it is referred to as just "TRAK", and the original working title has become meaningless.

According to their own admission, TRAK conforms to the international standard for architecture description ISO/IEC/IEEE 42010. Furthermore, TRAK mandates no development process or methodology.

TRAK is intended as a lightweight, pragmatic, and view-oriented framework for each kind of complex system. It has five architecture perspectives (Enterprise Perspective, Concept Perspective, Procurement Perspective, Solution Perspective, and Management Perspective) each of which contains a number of related views. In contrast to the relatively large amount of views in other popular frameworks of the defense domain such as MODAF (47 views) or DoDAF (52 views), defines TRAK only 22 views.

Due to the fact that TRAK is open-source, the whole documentation is hosted on *SourceForge* (http://trak.sourceforge.net), a web-based repository which acts as a centralized location for free and open-source projects, mainly for the software engineering domain.

16.3.8   European Space Agency Architectural Framework (ESA-AF)

Brief Description	
First year of publication:	Unknown
Developed, respectively published by	Telespazio VEGA, on behalf of European Space Agency (ESA)
Primary field of application	Space System of Systems
Development process or methodology	Based on TOGAF (ADM)
Meta model (Ontology)	ESA-AF Meta-Model, based on UPDM

The exact release date of this framework is not known. ESA-AF [44] is based on established frameworks, such as MODAF and TOGAF. Its meta model is based on UPDM (see Section 16.4). Due to the fact that MODAF emerged in the defense domain, and TOGAF was primarily developed for Enterprise IT systems and infrastructure in the business domain, both frameworks don't address issues that are crucial for the space domain. Therefore, these known frameworks and methodologies have been tailored and extended to satisfy specific needs that are characteristic for Space SoSE. Some of these special needs are as follows:

- *Regulation needs in a multinational environment:* In European space programs, usually many European member countries are involved. Each participating nation has its own national regulation regarding governmental policies, processes, and procedures. ESA-AF supports the SoS engineering process in coordinating and aligning these individual rules and policies so that the space program can be successful.

- *Space domain needs:* The framework supports an accurate representation of space domain concepts and their relationships. For example, these concepts can be represented by space-specific types for modeling and a large set of parameters that are unique within the space systems domain. Furthermore, ESA-AF supports the representation of programmatic and procurement activities, which are central in the European space context.

- *Easy to use by a large number of different stakeholders:* Space programs are multinational undertakings that involve a large number of actors with very different backgrounds regarding domain experience, their roles (politicians, officials, managers, researchers, spacecraft engineers, software engineers, and various other technicians), technical skills, and their culture. Successful communication on various abstraction levels is a key for the success of the whole program. The ESA-AF addresses this need and supports SoS engineers by maximizing the effectiveness and alignment of technical and strategic decisions.

ESA-AF was used to support the SoSE activities for GALILEO (the upcoming European Global Navigation Satellite System), Copernicus (formerly GMES, the Global Monitoring for Environment and Security), and the Space Situational Awareness (SSA) program, which is a system that warns about dangerous situations in the outer space.

16.3.9 An Architecture Template for Moderate Complex Systems

Not everyone in the Systems Engineering domain is confronted with the architecture of large system of systems or enterprises. Most systems engineers develop technical systems with a more moderate level of complexity. Most of the previously presented architecture frameworks are too large and too heavy-weighted for this purpose.

The working group on moderate complex systems (AG MkS), a joint working group of the Gesellschaft für Systems Engineering e.V. (GfSE, the German chapter of International Council on Systems Engineering (INCOSE)) and the Swiss Society of Systems Engineering (SSSE, the Swiss chapter of INCOSE), has developed a template for system architecture documentation especially designed for systems with a low to moderate level of complexity. This template has already been discussed in detail in Chapter 6.

Interestingly, this template cannot only be regarded as a simple skeletal structure for documentation. In the broadest sense, we could also speak of an architecture framework for moderate complex systems. Referring to the definition of architecture frameworks in ISO/IEC/IEEE 42010, even the architecture template for moderately complex systems satisfies most of the characteristics of a framework. It takes into account stakeholder needs, their concerns and viewpoints, describes the purpose of the system, and contains a chapter for architecture description with the help of views that should conform to the viewpoints.

Unlike the comprehensive architecture frameworks discussed above, the architecture template for moderately complex systems defines no meta model or process. It is a very lightweight template that prescribes only the bare essentials that should be included in a good architecture description.

16.4   THE UPDM STANDARD

Especially, modeling tool vendors are challenged to support a variety of DoDAF- and MODAF-conformant framework adaptations (e.g., NAF) that have been developed during the last years to meet the unique needs of several domains and/or nationalities. The *Unified Profile for DoDAF/MODAF* (UPDM) is an Object Management Group (OMG) standard that consolidates the meta model of both frameworks and solves interoperability issues between tools. Its current version UPDM 2.1 was published in August 2013.

The UPDM profile is the result of an initiative to develop a UML-conformant modeling standard that supports both DoDAF and MODAF. As a result, UPDM 2.1 conforms with DoDAF 2.02 Level 3, that is, it has physical data model conformance with DoDAF 2.02. It is expected that many other view-oriented architecture frameworks are largely supported that either have DoDAF- or MODAF-based meta models as their basis.

The package-oriented language architecture of UPDM provides two compliance levels:

- **UPDM 2.1 Compliance Level 0** is an implementation of UPDM extending the UML 2 meta model and importing several stereotypes from the *Service-oriented architecture Modeling Language* (SoaML) profile. For instance, some of the imported stereotypes from SoaML are Capability, ServiceInterface, Service, and Participant. Service-Oriented Architecture (SOA) is an architectural paradigm for defining how people, organizations, and systems provide and use services to achieve results and to develop capabilities. Both DoDAF and MODAF contain a service-oriented view, respectively viewpoint of the enterprise, for example, the Services Viewpoint in DoDAF. Thus, the imported SoaML stereotypes are used to support the modeling of these views, respectively viewpoints.
- **UPDM 2.1 Compliance Level 1** is an implementation of UPDM that includes everything in Level 0 and imports the whole SysML profile. In other words, this implementation can be seamlessly taken forward into SysML modeling. With this compliance level, a system of systems engineer is capable to establish a seamless integration with full traceability in the modeling, starting with the whole enterprise architecture (visions, operational goals, services, organizations), down to the concrete systems (SysML) and the software (UML).

16.5   WHAT TO DO WHEN WE COME IN TOUCH WITH ARCHITECTURE FRAMEWORKS

As systems engineers, we can come in touch with architecture frameworks in different ways.

One possibility is that we need to do the systems engineering for just one component system of a system of systems compound. If this is the case, it may be that we are not even aware about it. For example, in the defense domain it may be that a governmental procurement authority awards a contract for a system development to your organization without that it's obvious for you that the system is intended for an integration into a SoS. The probability is very high that the requirements which are formulated by the contracting authority are derived from an overlying enterprise architecture, respectively from a planned system of systems. If that is the case, then it would be very valuable to establish a seamless traceability from a possibly existing system of systems model to the model of your component system. This is particularly well possible if both the SoS model, as well as the model of the component system are based on the same modeling concepts, that is, both use the UML or SysML as their core language.

In much rarer cases, a systems engineer comes in the position to develop a system of systems as part of an enterprise architecture. As we've seen in this chapter, architecture frameworks are quite different from each other, both in goals and in approach. The huge amount of architecture frameworks available is good news and bad. It's bad, because it increases the difficulty to choose the right one for the current application that suits best. On the other hand, this variety is a great chance, because often the best choice is to blend together different frameworks in a way that works best for the current application. No matter which framework you choose, consider that the major goal of an enterprise architecture is to deliver real business value as quickly as possible to its stakeholders.

16.6 CONCLUSION

This chapter provided a broad introduction to the field of architecture frameworks. Due to the complexity of this topic, as well as by the enormous variety of available architecture frameworks, a deeper discussion is out of scope of this book.

Perhaps already the museum, which is the organization as well as the cultural institution of our case study in this book, can be considered as an enterprise and a system of systems that could be planned and developed with the help of an architecture framework.

But imagine this: many museums all over the world collaborate with each other and wanted to offer an incredible experience that cannot be realized by a physical visit. All together want to offer a global virtual museum tour, independent of time and space. For the virtual visitors, this means that they can watch all the paintings of Pablo Picasso that are distributed across the participating museums in the form of an artist-specific tour. Or they can look at various works of a certain period of art, regardless of whether they are physically located in an exhibition in New York, London, Paris, or Berlin.

Technical systems are needed for such a project that controls and coordinates the movement, and the distribution of the video streams of all the robots in the participating museums. It seems obvious that such kind of enterprise could be well developed and planned with the help of an architecture framework.

Chapter 17

Cross-Cutting Concerns

Cross-cutting concerns are those concerns that are relevant across different levels, layers, and perspectives of the architecture. There are very different concerns that have this nature. In this chapter, we will discuss some of the major ones. Each of the following sections is about one of them. Finally, trade studies and budgets will be discussed as a means of handling cross-cutting concerns.

17.1 THE GAME-WINNING NONFUNCTIONAL ASPECTS

Fried and Hansson write: 'things like speed, simplicity, ease of use and clarity are our focus. Those are timeless desires. People aren't going to wake up in 10 years and say, "Man, I wish software was harder to use"' ([42], p. 85). This statement stresses the importance of nonfunctional aspects, here: speed and ease of use.

Often nonfunctional requirements are cross-cutting concerns, because many parts of the system can spoil satisfying them. The reaction speed of a system can be spoiled by both a poor-layered architecture and an unsatisfactory product architecture. A noncompliance

Model-Based System Architecture, First Edition.
Tim Weilkiens, Jesko G. Lamm, Stephan Roth, and Markus Walker.
© 2016 John Wiley & Sons, Inc. Published 2016 by John Wiley & Sons, Inc.

with requirements about total mass of the system can result from a single part of the system being too heavy.

The system architecture description should describe how nonfunctional requirements are mapped to the different system elements. In the case of requirements about linear superposing parameters like mass or storage consumption it may be sufficient to make budgets of maximum parameter values for different system elements (see Section 17.5).

Important nonfunctional aspects are the so-called "-ilities" (which is a term from the "Terms and definitions" section of the INCOSE Systems Engineering Handbook [56]): availability, maintainability, manufacturability, reliability, supportability, usability, and so on. Even if not ending with "ility," safety should also be on the list of important nonfunctional aspects.

Nonfunctional aspects are often cross-cutting, because we may need to optimize in different perspectives for satisfying such requirements. For example:

- In the functional perspective, one may for example drive the optimization of power consumption by asking: Do we really need this functionality? Can we make it simpler?
- In the physical perspective, one can drive power optimizations by asking: Can we optimize the physical implementation of this power-consuming circuit?

17.2 HUMAN SYSTEM INTERACTION AND HUMAN FACTORS ENGINEERING

When we introduced the system context in Section 8.2, we placed users like the "Virtual Museum Tourist" outside the system boundary. This is necessary in order to define the correct scope of development artifacts, which ensures that the system and not the user is developed. Since the user is not part of the system, it is clear that the user interface has to be part of the system requirements that are fed into the systems architecting process. If the user was part of the system, then the user would become just another system element that the system architect can decide to move around in the system or even eliminate from it. In the case of our VMT example system, this would of course be nonsense. In systems involving operators, it can indeed be an option to move or even suppress the operator. Think of the systems involved in asking questions about your electricity provider's bill by calling this company. In the last millennium, there were times during which one needed to talk to a human operator for establishing the connection between the own home phone and the phone system of the electricity company. Today, we just dial a number and get connected — maybe again with a human operator who works at the electricity company. A trend in this area is to move the operator jobs to countries with low salaries or to spread them across time zones in order to offer services 24 hours a day without having operators do night shifts. Some companies even answer easy questions like "have I paid my last bill" without any human interaction on their side, by means of a computer system that can synthesize voice and asks the caller to navigate by pressing keys on the phone.

Humans can thus be either part of a system or outside the system. In both cases, the human system interaction has to be considered. Especially, in cases in which humans are outside the system boundary, like in the VMT case, we see a risk that the user is not getting enough attention. But the user is part of the system context and therefore in focus for the system architect, even if outside the system boundary.

We should thus consider all humans inside or in interaction with the system during its full lifecycle (so explicitly including, e.g., production and maintenance personnel). Human factors engineering is about fitting the system to these people. This involves, for example, ensuring that people will be able to use the system properly, but also caring about their safety. It can also include approaches like "design for user experience" Hassenzahl [51], with a focus on the emotions of the people in contact with the system.

Lockett and Powers [90] point out that human factors engineering extends "beyond the application of common sense to design" (p. 493). It is thus a discipline to explicitly account for in work breakdown structures and plans. Special expertise in this field of engineering is required. It would be beyond the scope of this book to provide more details on this special but very important field.

17.3   RISK MANAGEMENT

Again, risks are cross-cutting concerns, because the system and each system element down to parts can cause risks.

One has to distinguish product risk from project risk. While *product risks* are potential problems the product can cause during its use, *project risks* are potential problems in reaching the goals of the project. Project risks can be, for example, potential problems in meeting deadlines, staying in budget or ensuring that the development of the product is feasible at all.

In the scope of systems architecting, both kinds of risks are relevant:

- System requirements can state measures for reducing product risk, which have to be transformed into a solution via systems architecting.
- During their work, system architects may discover project risks, which they should feed into risk management.
- System architects should participate in proactive activities that aim at monitoring and managing product and project risks.

The INCOSE Systems Engineering Handbook [56] defines a risk management process that provides the risk strategy, the risk profile, and the risk report for a project. During the process, risks are *treated*; this means that actions are planned in case a risk becomes unacceptable. In order to do so, risks need to be analyzed by investigating the likelihood of the occurrence of an undesirable event as well as the consequences in case of occurrence.

Depending on various factors like the kind of products to be developed, the target markets or the kind of organization developing the product, there may be regulations in place that need to be taken into account when doing risk management. We do not go into details on these, since they differ from case to case. We recommend to anyone responsible for risk management to get to know the applicable regulations.

Finally, it has to be stated that looking at risks only may be a too narrow point of view. Forsberg et al. [41] write about "opportunities and their risk." They point out that risks arise from pursuing new opportunities and that the awareness of this fact allows for looking at the balance between risks and opportunities and optimizing it.

17.4 TRADE STUDIES

The INCOSE Systems Engineering Handbook [56] introduces trade studies as "an objective foundation for selecting one of two or more alternative approaches to solve an engineering problem." In the context of this book, trade studies are recommended, for example, for choosing the right product architecture during functional-to-physical mapping.

During a trade study, criteria will be defined for deciding on the solution. Then different solution options will be identified, where some may already be excluded in an early stage due to not meeting requirements toward the system. Afterward, decision-making is applied in order to decide for the solution to pursue. This can be made via a simple pro/contra analysis (which Ullmann [138] calls "Franklin's method" based on a quotation of a letter from Benjamin Franklin in which step-by-step guidance for a semiformal pro/contra analysis is given). Also more formal approaches like decision trees (e.g., Skinner [127]) can be considered.

In the end what counts is to achieve sufficient certainty to select the solution close enough to the optimum. The decision with its related rationale needs also to convince relevant stakeholders. Skinner [127] points out the importance of a communication plan for building trust in the course of actions to be taken during decision-making. Emes [35] recommends to make the selection of stakeholders based on the scope of the decision.

17.5   BUDGETS

Unlike the budget we give children before they enter a toy store, *budgets* in the following are not about money. This section is about budgeting certain properties of the system per system element in order to be able to meet a nonfunctional requirement whose scope is the complete system.

Budgets are needed if multiple system elements contribute to the same property of the system as whole. Examples of such properties are current consumption, heat dissipation, memory use, and mass. One can easily verify that each of these will be made up by contributions of all parts in the system having the corresponding property.

In order to make a budget, one will need to know the relationship between the property values of the system elements and the property value of the complete system. In the example of mass, the relationship is given by the sum operator: the total mass is the sum of masses of all parts. Already when considering storage, the equation may become more complex because overhead for the organization of data structures or the nonlinear effects of data compression algorithms may have to be taken into account.

As soon as one can compute or estimate relationships between the involved values, a budget for each of the parts can be made. For example, the maximum mass of each part can be derived from the specified maximum mass of the system. It is important to pay attention to tolerances when making the budget. Properties like mass are usually specified via a nominal value and a tolerance range. In order to be sure that a maximum mass requirement is met, one has to make a budget that states the maximum value inside the tolerance range instead of the nominal mass. Or more general: budgets are usually based on a worst-case scenario. However, it makes sense in certain cases to exploit statistical effects. In the given example, one can build the system by combining parts with lower than nominal mass and parts with higher than nominal mass such that the deviations from nominal partly compensate for each other.

Chapter **18**

Architecture Assessment

Experienced architects have a certain kind of a gut feeling if a design is good or bad. Many of the architects' decisions must be made early in the development phase on the basis of assumptions and can, if they have been bad, bring the project to failure. It is often difficult to roll back or revise architecture decisions and their consequences. To put the important architecture decisions on a profound basis you should do an assessment of your architecture.

Architecture assessment methods assess the quality of an architecture. It is not an analysis of the architecture that could be automated and leads to a set of objective performance indicators. We recommend an architecture assessment method that is based on a structured communication process to incorporate all relevant stakeholders and viewpoints. The assessment method cannot replace the architects' decisions. It focuses on the main important aspects, incorporates stakeholders to strengthen the decisions and provides a replicable documentation of the arguments for and against the decisions.

Model-Based System Architecture, First Edition.
Tim Weilkiens, Jesko G. Lamm, Stephan Roth, and Markus Walker.
© 2016 John Wiley & Sons, Inc. Published 2016 by John Wiley & Sons, Inc.

The Architecture Tradeoff Analysis Method[SM] (ATAM[SM])[1] developed by the Software Engineering Institute (SEI) at the Carnegie Mellon University is a common architecture assessment method in software engineering [72]. Since it does not focus on specific software technologies, it could also be applied to the systems engineering discipline (e.g., Ref. [39]). The SEI has also developed a special variant of ATAM for Systems Engineering that in particular addresses system of systems.

ATAM leads to an explicit consideration of the system objectives, related quality attributes, architectural approaches, and decisions.

Typical quality attributes of an architecture are for instance performance, reliability, availability, security, producibility, disposability, and modifiability. They are tightly coupled to the physical architecture. Some of the quality attributes are also relevant for the functional

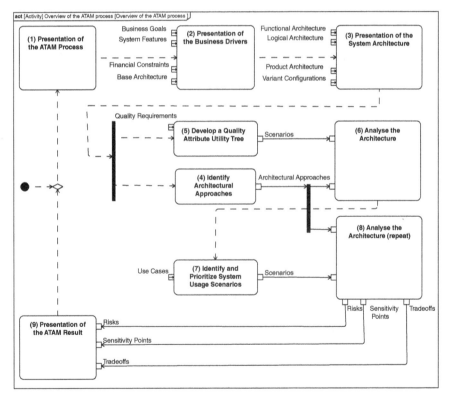

Figure 18.1. Overview ATAM [72].

---

[1] Architecture Tradeoff Analysis Method and ATAM are service marks of Carnegie Mellon University.

architecture (Section 14.10). For instance, performance attributes about the duration of a function to perform the result.

The ATAM process can be performed in early phases of the system development project. Since you don't need the real system to assess the architecture, it can be already done with early versions of the architecture documentation. That uncovers potential risks and sensitivity points early enough to adapt the architecture if necessary. It also highlights tradeoff points that forces the stakeholders to prioritize their requirements. For instance to answer the question if it is more important that the system meets the energy consumption or the performance requirements which are contradictory requirements.

ATAM mainly considers three aspects: the goals of the system and the business around the system, the quality requirements that meets the goals, and finally the architecture that satisfies the quality requirements. ATAM has nine process steps (Figure 18.1):

1. Present the ATAM process
2. Present the business drivers
3. Present the System Architecture
4. Identify Architecture Approaches
5. Develop a Quality Attribute Utility Tree
6. Analyze the Architecture
7. Identify and Prioritize System Usage Scenarios
8. Repeat step 6: Analyze the Architecture decisions based on the outcome of step 7
9. Present the ATAM Result.

In the remainder of this chapter, we briefly describe the ATAM process to understand the concepts. We've adapted the ATAM concepts to our architect toolbox and integrated artifacts like the base, logical, and product architecture into the ATAM process. For a more comprehensive documentation of ATAM, please refer to the report from the SEI [72] or the book "Software Architecture in Practice" by Bass et al. [10].

The first three steps are about presenting the ATAM process itself, the goals of the business and the proposed architecture to selected stakeholders of the system. That clearly shows that architecture is a lot about communication and not limited to pure engineering tasks. See Section 20.1 about communication skills for system architects. After the presentation of the ATAM process, typically presented by the lead of the architecture assessment team, the project manager presents the business drivers of the system. These are the most important features (functional requirements), the business goals, financial and other constraints, and other important aspects for the business system context. Figure 18.2 shows on the left side the main four objectives of the system of interest. These objectives address the concerns of the operator of the system, that is, a museum. The business goals address the concerns of the developer and vendor of the VMT. A business goal of the VMT is that the vendor will be the market leader for virtual museum systems. Financial constraints are the maximum costs of the development. Other constraints are for instance the base architecture (Section 7.2) and the organization of the development project (staff, locations, and processes). Overall each presentation should not last longer than an hour.

A system architect gives a presentation of the architecture. It should cover the different architecture types, that is, the functional, logical and product architecture, their relationships and the main elements of the architecture. In addition, the main driving requirements of the architecture, for instance performance requirements like a low mass of the

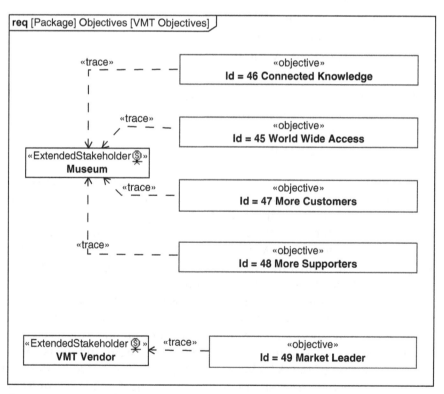

Figure 18.2. VMT objectives.

system, are presented and some main use case scenarios are given. The system model is a perfect source for the presentation. Specific diagrams could be created to address the concerns of the audience. All diagrams are views on the same model. It is the same model that is used for the development of the system. Presentation and real world are identical. Again that presentation should last round about one hour.

The fourth step identifies the architecture approaches. In the terminology of ATAM architectural approaches are a set of architectural decisions. An architectural decision could belong to an architectural style. A style provides a vocabulary for the architectural elements and connections and constraints how they could be connected. It is not a concrete pattern that solves a concrete problem. A pattern could be part of a style. An example for architectural style in software engineering is a client/server or peer-to-peer architecture. The term architectural style is defined in Ref. [43]. Although its origin is the software engineering discipline, it is also valid, usable, and valuable in the systems engineering discipline.

The final step before starting the analysis is to identify and prioritize the most important quality attributes of the system by building a quality attributes utility tree. Quality requirements from the business perspective are often too fuzzy. For instance, "The system must consume little energy" or "The system must be easy to use." The utility tree is a tool to breakdown high-level quality requirements to concrete scenarios. Figure 18.3 shows a utility tree of the VMT. The first level is a list of quality requirement categories like performance, availability, security, and safety. The leaves are concrete scenarios that could be prioritized. Kazman et al. propose a two-dimensional prioritization [72], where the first dimension states how important the success of the scenario is for the system and the second dimension state the criticality of

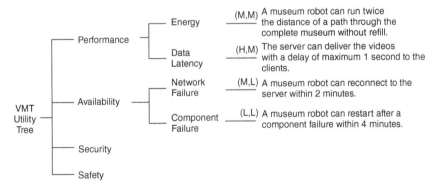

Figure 18.3. Excerpt from the quality attributes utility tree for the VMT.

the successful achievement of the scenario. Simple stages for a priority are High (H), Medium (M), and Low (L). During this step, you typically identify new requirements or requirements that need to be changed. You should involve the requirements engineers at this point. It is another example for the close collaboration of requirements engineers and system architects. See also Sections 7.1 and 7.2.

The central step of the ATAM process is the sixth (and eighth) step. It analyzes the architecture decisions—identified in step 4—and reports about the risks, sensitivity points and tradeoffs of each decision. The utility tree - developed in step 5 - and the included prioritized scenarios are the entry point of the analysis. Starting with the high prioritized scenarios each scenario is linked with the architectural approaches that are relevant to achieve the scenario. Output of step 5 is a list of the architectural approaches together with a list of risks, sensitivity points, tradeoffs, and questions.

The utility tree is driven by the quality requirements of the system. In the seventh step of ATAM, we brainstorm scenarios that describe the usage of the system and scenarios that describe potential changes of the system. A good source for the usage scenarios is the list of use cases identified during requirements analysis of the system (Section 8.3). If you identify further usage scenarios you've probably found a new use case and must integrate the new information into the requirements and use case analysis. The change scenarios are used to test the capability of the system for future changes. Kazman et al. describes two kinds of change scenarios: growth scenarios and exploratory scenarios [72]. A growth scenario describes expected changes in the near future, for instance, an increased number of tourists who uses the VMT or to combine VMTs to provide a VMT across several museums—a huge virtual museum. An exploratory scenario is an extreme growth scenario. An example of such a scenario could be that the number of simultaneous users of the VMT grows exorbitantly. They are not expected to occur, but help to identify sensitivity points of the architecture.

The eighth step repeats the analysis step 6 with the outcome of step 7. Finally, step 9 is a presentation of the overall outcome of ATAM to the stakeholders. Of course, the outcome of ATAM could result changes of the architectural decisions, quality requirements, and even the business goals. After applying those changes you can restart the ATAM process, and so on.

We recommend not to consider the ATAM process as a strict workflow instruction. Keep the principles of ATAM and adapt the process to your specific needs.

Chapter **19**

Making It Work in the Organization

19.1 OVERVIEW

None of the authors has so far seen an organization that was founded with a system architecture department in place. Most organizations discover over time that development of their products or other systems of their interest require systems architecting for instance for the sake of efficiency and quality or to be competitive. Introducing system architecture in an organization is an organizational change process, and it is not an easy one: while some organizational changes aim at making the same work more efficient, the introduction of systems architecting in an organization may lead to a different kind of work, which requires a different kind of thinking and acting.

Even in case systems architecting is well established in an organization, it has to be kept operational during future organizational changes. A continuous organizational roll-out of the methods and the mindset of systems architecting is required from those who are convinced that systems architecting is one of the enablers for success—no matter how much has been established before. In organizations that are used to systems architecting, this continuous process ensures a reminder for most and an introduction for new employees. In organizations that have not

Model-Based System Architecture, First Edition.
Tim Weilkiens, Jesko G. Lamm, Stephan Roth, and Markus Walker.
© 2016 John Wiley & Sons, Inc. Published 2016 by John Wiley & Sons, Inc.

been doing it before, the process ensures that the organization learns to explicitly architect its systems of interest.

It does not matter whether systems architecting comes top-down from management or bottom-up from engineers who like to establish a new mindset: there needs to be continuous efforts to make and keep systems architecting work in the organization. In this chapter, we first discuss the organizational structure around systems architecting and then provide some recipes from our own experience for making systems architecting work in the organization. One can usually not assume that the people driving system architecture in an organization are in the position to perform an organizational change project. This chapter therefore does not discuss the management of an organizational change. It focuses on approaches that can help creating buy-in for systems architecting on a peer-to-peer level. Even if you are currently driving an organizational change, you will still need to do this, in order to make the stakeholders enthusiastic about the value of systems architecting.

19.2 ORGANIZATIONAL STRUCTURE FOR SYSTEMS ARCHITECTING

The *organizational structure* or *organization structure* means the "formal patterns of how people and jobs are grouped in an organization" (definition according to the glossary in the book "Organizations" by Gibson et al. [45]). We can represent it with an organizational chart. Blanchard [13] presents different organizational charts for showing potential organizational structures that support systems engineering via dedicated entities in the organization.

Since our scope is limited to systems architecting, the organizational charts we discuss have their focus on systems architecting. One should consider grouping different systems engineering disciplines into one systems engineering entity, which is however not shown here because it would be beyond this book's scope. In other words, where we write "Systems Architecting" one can as well read "Systems Engineering" and imagine "Systems Architecting" as one entity inside it.

Inspired by Blanchard's mentioned organizational charts, we show some hypothetical organizational structures that host systems architecting work in Figure 19.1. The names of the different levels and of some of the organizational units are taken over from Blanchard [13]. We expect these names as well as the whole shape of the organizational

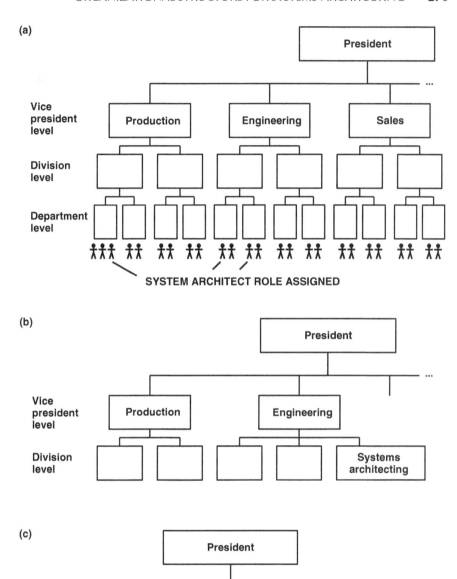

Figure 19.1. Examples showing different alternatives for anchoring systems architecting in the organization, where option "(c)" is purely hypothetical.

chart to vary from organization to organization and they should therefore be considered as examples only. We have chosen examples that represent the traditional hierarchically divided kind of an organizational structure, because we assume that it is well known. In reality, we may find different kinds of organizational structures. Some examples are summarized in Gray's and Vander Wal's book "The Connected Company" [47].

All three different organizational charts in Figure 19.1 aim at ensuring that the system architect role we defined in Chapter 11 is assigned to some people working inside the organization:

- In Figure 19.1(a) the role is assigned freely to employees with the appropriate skills, where needed. These employees should of course be trained as system architects (see Section 11.6) and should be organized as a system architecture team (see Section 11.3), which is not shown in the organizational chart. The people with the system architect role form a community of practice (see Wenger [149]).

- In Figure 19.1(b) there is a dedicated organizational entity for systems architecting. It resides inside the entity "Engineering."

- In Figure 19.1(c) a dedicated entity for systems architecting directly reports to the president. It is there as a hypothetical possibility to be used in the discussion of the advantages and disadvantages which will follow.

In order to find the optimum anchoring of systems architecting in the organization, the different options from Figure 19.1 have to be compared with each other. This is what we will do in the following; but first we like to point out that also mixtures between the different options need to be considered in order to increase the likelihood that a good solution is found. We will first discuss which level in the organization to choose for systems architecting before we compare the pure role approach according to Figure 19.1(a) with the ones based on organizational entities according to Figure 19.1(b) and (c). In the end, we will discuss mixtures between the different options.

When we consider that the system architect has to optimize the system in a holistic way, then an organizational entity that is responsible for systems architecting should in a straightforward approach be placed on an equal level or on top of those entities that define and implement the different subsystems. Based on this consideration only, placing systems architecting on the same level as the different engineering disciplines like shown in Figure 19.1(b) is probably a good option. However, let us revisit a finding from Chapter 10: the system architect should not only be concerned with the system of interest but also with its enabling systems. Consider production systems like automata used during assembly, programming, and production testing of the final products as one example of enabling systems. In the organizational charts in Figure 19.1, they may be in the responsibility of the entity "Production," so outside the "Engineering" divisions. In organizations in which interactions are difficult across high distances in the organizational chart, one would thus need to place systems architecting at least on the same level as the "Production" entity, which would result in the variant shown in Figure 19.1(c). The "Production" entity was just an example here. There are many more such entities with relevance for the system architect that can only be reached directly on the level shown in Figure 19.1(c). Examples are the information technology entity and the one(s) ensuring maintenance and after-sales support of the products.

We can now ask ourselves if a systems architecting entity directly reporting to the president like shown in Figure 19.1(c) is reasonable. In many cases, it is probably not. In organizations in which interactions are difficult across high distances in the organizational chart, the system architects might need to be placed on a level where they are close to the very important engineering stakeholders, especially developers doing the actual implementation work. In organizations in which hierarchical distances do not matter, the position of a system architecture entity does not matter so much anyway.

It should be questioned whether at all systems architecting needs to be represented by an organizational entity or whether it can be implemented by assigning the system architect role to different people in existing entities, like shown in Figure 19.1(a). Before discussing this, let us revisit this role assignment approach and add some details to it: the idea is to have people with an engineering background and existing assignments inside the engineering disciplines of the organization do systems architecting. They can be trained as systems architects. They then take the system architect role and use parts of their work time for systems architecting work. The rest of the time remains available for their other engineering work. The people with the system architect role should be joined in a system architecture team, and the leader of this team has to be defined. The clear advantage of this approach is the direct link it creates between the engineering work and the systems architecting work. It ensures that system architects are aware of the current situation in the engineering disciplines and that they can easily communicate the system architecture to the engineering disciplines in a peer-to-peer fashion. Of course, there are also some disadvantages to the approach. Here is a comparison:

- Advantages of assigning system architecting to individuals in the organization via the system architect role according to Figure 19.1(a):
  - The systems architects are in close contact with the engineering disciplines.
    * This eases the maintenance of communication and networking between systems architecting and engineering.
    * The system architects stay in touch with reality. In this context, it is noteworthy that Muller [98] sees system architects "drift away from reality" as a consequence of working in abstractions too long.
    * The system architects stay up-to-date with new evolutions in the different fields of engineering.
    * The system architects maintain credibility in the engineering disciplines because they are recognized as peers.
  - The organizational structure does not need to be changed. This is an advantage particularly if systems engineering is new for the organization, which implies that there is no knowledge or experience with systems architecting that would allow for deriving the appropriate organizational structure.
  - The efficiency of systems architecting is more invariant to organizational change than in the other shown options because the system architects are used to work in a network across the organization.
- Advantages of creating a dedicated systems architecting entity in the organization according to Figure 19.1(b) and (c):
  - There is visibility of the systems architecting tasks in the organization.
  - It is easy to understand who has to handle the systems architecting tasks.
  - Work-load in systems architecting can be made more predictable over time (see also Section 10.6).
  - The people in the systems architecting entity can fill their work days with systems architecting tasks, without conflicts with engineering tasks.

> – A person in the systems architecting entity can become an expert in systems architecting without a conflict with the need to stay expert in an engineering discipline.

The last two subitems of options (b) and (c) can also be the case with option (a), if the people with the system architect role are allowed to use their time fully in that role.

The above comparison shows that there is not the ideal solution. One needs to optimize the setup based on the current situation in the organization. The optimum setup may be different after reaching a certain maturity in systems architecting than during the first time an organization consciously does systems architecting.

As mentioned above, it is also possible to mix the different options:

- One full-time architect or a few ones can work in a dedicated entity in the organization and run architecture team work with individuals from the engineering disciplines to whom the system architect role has been assigned.
- A dedicated entity according to Figure 19.1(b) can host the system architects that work on the system of interest, whereas several individuals who work on enabling systems get the system architect role.

The conclusion is thus that there is not the ideal organizational structure anchoring systems architecting in an organization. The organization needs to be designed individually, based on the specifics of the business, the enabling systems, the maturity of systems architecting in the organization and many more factors. No matter whether there is a dedicated entity for systems architecting it seems to be a good idea to have direct links to people inside the entities of the organization that are concerned with engineering the system of interest and developing or purchasing and operating the enabling systems. Whether this is achieved by assigning the system architect role to people inside those entities or by establishing a good network for the systems architects has to be determined on a case-by-case basis.

19.3   RECIPES FROM THE AUTHORS' EXPERIENCE

The following recipes aim at making systems architecting work in an organization that has not consciously done systems architecting before on the one hand, and on the other hand they should help maintaining and improving systems architecting in organizations in which it has already reached a certain maturity.

19.3.1   Be Humble

We have seen that there are many stakeholders of system architecture (Chapter 10). With the right mixture of luck and professional skill, some of them will enable no less than the success of the system: the requirements engineers will have described the right product for being successful, the engineering disciplines will have properly implemented the subsystems, and the verification people will have observed everything that needs to be observed before a release to the market, well before the release.

And the system architects? Are they contributing to the success of the system? Of course. A system with a really bad architecture is not likely to have success, whereas there are ways in which a good architecture contributes to success (as it has been discussed in Chapter 3). It is thus allowed for the system architect to be proud of having worked on successful systems. However, the system architect should be humble and acknowledge the contribution of the stakeholders as the ones that had the direct impact on the system's success. Otherwise, the system architects might be perceived as parasites who are living on other people's successes. This may be an impediment to the commitment to systems architecting in the organization.

19.3.2   Appraise the Stakeholders

In order to be able to stay humble as recommended before, the system architect may consider that the stakeholders of the system architecture are the real heroes, whereas the system architect is the one coordinating between their disciplines and ensuring that the solution is holistically seen the right one. The system architect should show the stakeholders that he is aware of the importance of their contribution. On this basis, it is easier to show the importance of the system architect's own contribution to successful system realization. Ideally, the system architect together with the stakeholders will be seen as the winning team whose members can only reach success efficiently if working together as a team.

19.3.3 Care About Organizational Interfaces

Important organizational interfaces for the system architect are the ones to the architecture stakeholders. The system architect should periodically assess whether these interfaces are well established, in other words: that the network is alive. This is independent from the approach of anchoring systems architecting in the organization, so it is applicable to both people working in a dedicated organizational entity for systems architecting and people to whom the system architect role has been assigned.

The periodic assessment of organizational interfaces can be carried out according to the following procedure:

- Consider all possible entities you know in and around your organization and rate whether the system architects' interface to that entity is important.
- For the important interfaces from the previous step: rate how well the interface is established.
- Where there is a mismatch between importance and strength of the interface, there is a need to act. Assess whether the need is immediate; then create a map as shown in Figure 19.2.

When there is a need to act, define a strategy for improving the organizational interface. Here is a nonexhaustive list of proposals how this may be achieved:

- Prioritize systems architecting work for activities involving the stakeholders with whom the interface needs to be improved.
- Invite the given stakeholders for sessions of the system architecture team, and ensure that the agenda items to cover during these sessions are relevant for them.
- Extend the scope of the tasks to be done for the given stakeholders: ensure that more detailed work than usually is done in order to reach a better common understanding with the stakeholders.

19.3.4 Show That It Was Always There

As given in Chapter 4, each system has a system architecture. This is true even if there has never been a dedicated systems architecting task during the realization of the system. The system architecture may thus have been established implicitly while developing the system. When stakeholders had to learn cooperation with system architects then we often

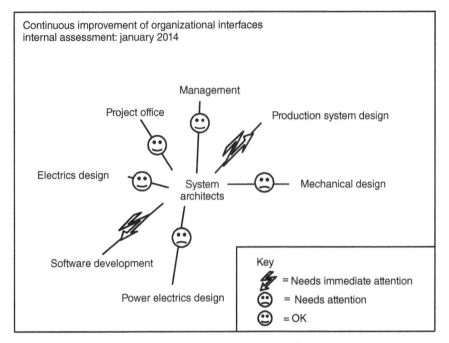

Figure 19.2. Example showing the rating of how well organizational interfaces are established. The organizational information is freely invented.

noticed they were concerned about additional work-load. In this case, it may be helpful to explain that the stakeholders themselves are the ones implicitly doing the systems architecting tasks in a world without system architects. The systems architecting work can then be recognized as something that has always been done. The difference with proper systems architecting in place is that the corresponding work is done explicitly and that each system architect helps getting it done. Instead of imposing additional work-load, a system architect thus helps finishing the work.

19.3.5 Lead by Good Example

It is own behavior that will drive change. This is also the case in establishing and sustaining systems architecting in an organization. The system architects should comply with their own paradigms if others are expected to follow them.

For example, if it has been decided to follow the model-based approach and to make the model the single source of truth, then the system architect should never archive a temporary document "just for now until we have time to put it into the model." Of course, we have seen in Section 9.11.2 that it may be possible to work on printed posters for the sake of conducting efficient workshops and to feed information back into the model later. It has also been pointed out by Scott Ambler [8] that some models do not even need to go to the archives. We would not like to encourage anyone to start pointless activities just because the direction they follow may have some superficial resemblance of a given paradigm. We rather encourage you to look for the good examples that show the value of chosen way of systems architecting. In the given example, one may say: "in this workshop we have made a lot of modifications of the model on a poster. We will put them in the model later. However, one key result of this workshop is to add a new operation to the Museum Robot's service interface. We will now open the modeling tool and insert it together so that it becomes visible for all model users right now." If it then turns out that the modeling tool is too cumbersome to use in front of a group of people, then the homework has not been done.

In order to lead by the good example, one needs to do homework. If you are *uncomfortable* with the methods, processes, or tools that you are promoting, then you should assume that they will be *unusable* for others. So better come prepared and make things work well for yourself before you ask anyone else to follow you.

19.3.6 Collect Success Stories and Share Them When Appropriate

Systems architecting generates value indirectly: it enables the organization to achieve certain benefits, like it has been discussed in Chapter 3. Therefore, it is necessary to show how systems architecting contributed to the success of the organization. This can be achieved by telling success stories about systems architecting at the appropriate moment. One of the authors has just recently used a four-year-old success story to get the budget for a new systems architecting activity approved, because the old success story matched the current situation so well that it became hot news again.

We have experienced that it is problematic to use stories like "Company XYZ published a report proving that they have doubled productivity through systems architecting." No one can prove that the preconditions for company XYZ were the same as they are in your organization and in your current situation. It is much more appealing to use success stories from the own organization, because people will remember the case and will be able to confirm the story intuitively, even if the data backing it up is poor compared to published reports. But the only way to have these stories ready for telling them when needed is to collect them when success is achieved. Build your own collection of success stories and remember to update it on each new success.

What can go into these success stories? In Section 3.3, we already recommended to collect affirmative feedback. Such feedback can be archived and can be used later to tell a success story. But we strongly recommend to ask feedback providers for permission before quoting them in highly visible documentation. Apart from affirmative feedback you can use your own perception of successes as long as they are transparent for others. In case of doubt, have your success story reviewed by a stakeholder.

Success stories can be made as case reports: what was the problem, how was it solved, and how did the chosen systems architecting approach contribute to success considerably? Avoid collecting success stories that start with a problem like "The XYZ department had forgotten to analyze a dependency" and then tells the story of the heroes from systems architecting that saved the boat. Some people will not like the story (especially the ones from the XYZ department) and this may spoil the whole message. Good success stories are in a shape like "A typical system XYZ project takes N months, but in project ABC we needed two months less. We had applied the FAS method, triggered by the problem that a new requirement needed to be taken into account, without any prior knowledge in the organization on how to satisfy it. We chose the FAS method, because the problem space and the solution space needed to be analyzed in a structured way."

Ideally, you have a template ready for collecting success stories. One of the authors uses an empty slide show in the organization's corporate design. The template consists of a title "Success story about XYZ" and some instructions. The writer is supposed to fill in "XYZ," to produce a visually appealing case report that tells the success story and to continue the story in case the amplitude of the success or the evidence for the contribution of systems architecting to success increases over time. It is important to record the preconditions and the achievements. This makes future situations comparable with the one at hand. As a consequence, it can be judged in future whether learnings from a success story are applicable to a new situation.

Tell success stories when the situation is appropriate. This may be when you are asked to report status or when you like to get approval for a new activity or a budget. On the occasion of a status report, you may state the usual "activity XYZ finished within schedule and within budget" and then continue stating "but this time we got the feedback from the engineer John K. that he had a much better overview of his task right from the start of the activity, because the system architect had generated a special view from the model to visualize the context and the fundamental assumptions in an intuitive diagram." When asking for budget, you might state "Last time we applied this method we finished 2 months earlier than usual. This time we would like to invest into some tool support for the method."

19.3.7 Acknowledge That Infections Beat Dictated Roll-Out

Herrero [52] points out that changes can travel via networks and "infect" others. It is thus important to do networking in order to make changes toward better systems architecting happen.

System architects should not believe that the word "system architect" on their business card automatically gives them the authority to just dictate the valid system architecture. The definition of the system architecture results from collaboration with the stakeholders rather than from a one-way roll-out.

If it is possible to convince the stakeholders of the value of systems architecting while working with them, then the chances are high that commitment to systems architecting will spread like a virus. Of course, the system architect ideally needs the last word if the system as a whole would suffer from a solution that is proposed by one engineering discipline only. However, instead of optimizing solutions by means of having the last word, the system architect can use stakeholder-specific views to make stakeholders aware about the consequences of their work on the system as a whole. When the stakeholders understand the impact, it is usually not necessary anymore to convince them of a certain holistic solution. They will see the reasoning for it themselves and commit to it much more easily.

19.3.8 Assign the System Architect Role to Yourself

What to do if the organization is not at all ready for systems architecting and you think you are the only one realizing that it is necessary? In Chapter 11, we have seen that a system architect role can be assigned to an individual like a president role can be assigned to an actor who has never been elected. So why not assign the system architect role to yourself even if there is no role description for it in your organization.

Of course, you should not stand up and proclaim "I have made myself a system architect." You should rather use inspiration from the systems architecting approaches that you can for example read in this book and find out how you can make your daily work more efficient by using them. Before trying to spread anything in the organization, you should first prove to yourself what works well in your work area and what does not work. Once you have found the right approaches for your own work, your way of working has a good chance of "infecting" (Herrero [52]) others.

You should though be prepared for the day on which your organization decides to officially initiate systems architecting. If you have been pioneering enough good approaches and have achieved enough visibility then you may have the luck to be involved in the organizational change toward more systems architecting. For the case that you do not have that luck you should always be prepared that all the approaches you were test-driving as a self-nominated system architect may be overruled by an officially nominated system architect.

If you feel you are in the situation of driving systems architecting as a self-nominated pioneer, then you can seek assistance in the community. Consider becoming a member of the nearest INCOSE chapter (www.incose.org) if you are not yet one. Find out if there are local events maybe even at the place where you live or very close to it. You can use them to meet people who are or have been in the same situation as you. They can give you advice or help you find out that you are not alone with a problem. Eisenring et al. [32] report that some group members of a community working group in systems engineering were surprised how many coincidences they found between each group members' own challenges in daily business and other group members' experiences.

19.3.9 Be a Leader

Systems architects ensure that the definitions in the system architecture and the chosen architectural patterns are being followed. Whoever needs followers is ideally a leader. This is why leadership skills are important for a system architect (see Section 11.2.5). Leadership should not be confused with stubbornness in insisting on the own approach. Leadership is about behaviors that make others understand why it is necessary to contribute to systems architecting and to comply with the resulting definitions, based on the system architect's understanding of their situation and their point-of-view. It is about ensuring that it is possible to go into the right direction together while feeling relaxed about it.

Chapter **20**

Soft Skills

The fact that good communication and collaboration is crucial for a successful project is truly nothing new. It is the essence for any successful organization. A prosperous enterprise brings people together and creates an environment of good cooperation and appreciative togetherness. Especially, in times of ever-increasing complexity and globalization, humans and their collaboration are often the decisive factor for success or failure.

These are all truisms and known for decades. And also since decades, the creation of a collaborative environment for people is one of the major challenges for any organization. People are different regarding their personality traits, social graces, communication, language, personal habits, education, and friendliness. That means: in addition to the occupational requirements of a job, which are often called hard skills, there are other important skills a person should have that are related to the interaction and communication with others. These so-called soft skills are significantly influenced by a person's knowledge of human nature, socialization, feelings, emotions, empathy, personal insights, and cultural factors.

There are numerous publications and books about soft skills, discussing its importance and suggesting ways how to improve those skills.

Model-Based System Architecture, First Edition.
Tim Weilkiens, Jesko G. Lamm, Stephan Roth, and Markus Walker.
© 2016 John Wiley & Sons, Inc. Published 2016 by John Wiley & Sons, Inc.

It is not the purpose of this chapter to cover this topic to its full extent, but it provides a broad overview about some essential parts. In this chapter, you also will learn about some models from communication psychology and personality typology that are helpful to explain several interpersonal dealings and processes. First of all, these models are all wrong, because they greatly simplify the real complexity of these topics. It is basically not correct to pigeonhole individuals into certain categories. On the other hand, such models are useful to explain certain phenomena in a popular scientific way. Or as the British mathematician George E. P. Box (1919–2013) put it: "Remember that all models are wrong; the practical question is how wrong do they have to be to not be useful" [121].

20.1 IT'S ALL ABOUT COMMUNICATION

The internal communication within an organization refers mainly to the communications between members of that organization; in case of a company it is about employee communications. In addition, the internal communication is part of an organization's corporate identity. In this context, a good communication culture should develop a team spirit, a kind of shared identity between the members of the organization.

A vivid internal communication within a company is crucial for its success. This is especially true for development organizations employing knowledge workers. In today's world of ever-increasing complexity and dynamic markets with their unpredictable changes, it is of great significance for the existence of an organization to be able to react immediately on disruptive forces. Dealing with high dynamism, complexity, creativity, and productivity requires a high degree of communication on all levels.

While analyzing communication models, communication scientists basically consider two different categories of communication:

1. Communication as a transport and delivery process of a message between a sender and a receiver
2. Communication as a social activity within social structures.

These two categories are not to be understood as opposites, because the one includes always the other. The difference is basically in the focus of the communication process. The latter category considered communication not only as the sum of actions of individuals, but as a complex social process with emergent properties.

Basically, communication is fallible (error-prone). No matter how much effort you would put into your communication skills: this error-proneness cannot be fully eliminated. Generally, it is impossible to ascertain after a communication process has taken place whether a comprehension has taken place as intended.

This section highlights the importance of communication in systems engineering projects and discusses some topics that might be useful for the system architect's daily work.

20.1.1 Losses in communication

The Shannon–Weaver model, named after the American mathematician and electronic engineer Claude Elwood Shannon (1916–2001) and the American mathematician Warren Weaver (1894–1978), describes communication as a point-to-point-transmission of a message from a sender to a receiver [126]. It can be regarded as one of the simplest models and it's generally applied in various communication theories. In this model, influencing factors play an important role that were termed by Shannon and Weaver as "Noise". In the Shannon–Weaver model, this "Noise" primarily affects the transmission path between sender and receiver, that is, it deals with external disturbances only. Therefore, one of the often-heard criticisms of this model is that it is too simple to describe the real complexity of human communications.

As with message transfers in telecommunications engineering, communication between people is also never lossless. Just imagine that you should describe the beautiful beach from your last vacation in a way that someone gets an exact imagination about it. No matter how much effort you put into this, your counterpart will never have exactly the same image in mind as you.

The first loss could occur when you try to express your memories of the beach in language. Even if you can remember a lot of details, it is difficult to express every detail in language. "The beach sand was nearly white as snow." is not sufficient to describe the exact color of the sand. The sand color that the receiver of your information imagines depends on many factors. So it may be that she doesn't know snow and the metaphor therefore did not work.

The next loss could occur on the communication path between you (the sender) and your counterpart (the receiver). Just because you said something, it does not have to be arrived at the receiver. At this point, the so-called "Noise" described by Shannon and Weaver comes into play. External influences like disturbances or interferences can affect communications or even prevent it. For instance, there might be issues with the mobile phone network which directly affect the phone communication. However, in human communication such losses may not always be caused by technical issues. Maybe your communication partner didn't pay attention to your words, because she is distracted or not focused on you.

Another loss might occur when your counterpart didn't understand you. And that does not necessarily mean that such lack of understanding is caused by language difficulties. Of course, language and translation problems are often the cause of communication problems in international projects. Not infrequently, losses in communications in systems engineering projects are caused by a lack of understanding about domain-specific topics, or caused by disparate domain knowledge between different stakeholders (see Chapter 10). Stakeholders in a system development project have different concerns. Some of them are domain experts, others are users of the system, and others again are business people, but they are usually not experts in systems engineering. Probably one of the greatest challenges for the systems engineer is on the one hand to understand these stakeholders and their needs, on the other hand to explain them how the system of interest will work to satisfy their needs.

But even if there are little or no understanding problems: just because a message was understood and acknowledged by your partner, it does absolutely not mean that she agrees with its contents! Unfortunately, not everyone is aware about this important fact. Many think that it is merely sufficient to post an information to its recipients and everything is fine, but such behavior is for example not able to engender a commitment by the recipients of that information. Or to demonstrate it with an example: you too, dear reader, can read this book, but you may not agree with all of its contents. Although we, the authors, would be very pleased to get your feedback, you are not obliged to tell us your differing opinions and views.

20.1.2   The four sides of a message

There are a variety of different models about human communication. A well-known model that explains the complexity of communication is the so-called Four-Ears-Model, sometimes also called "The four sides of a message", developed by Prof. Dr. Friedemann Schulz von Thun [143], a German communications psychologist.

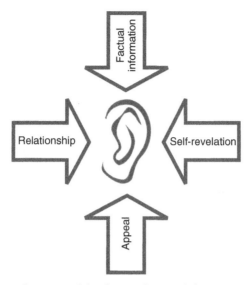

Figure 20.1. A visualization of the four-sides model.

Schulz von Thun postulates that each communication contains always four messages simultaneously and can therefore be received on four levels, sometimes also called channels: the factual information level, the appeal level, the relationship level, and the self-revelation level (see Figure 20.1).

- *The factual information level:* on this level, we send respectively receive the purely factual information. This level is about data, facts, and circumstances about a particular topic.
- *The relationship level:* on this level, the message conveys what kind of relationship the sender and the recipient have with one another.
- *The self-revelation level:* on this level, the message conveys information about the sender, something about what is going on in him/her.
- *The appeal level:* on this level, the message conveys a wish, a request, an advice, or an instruction; something of which the sender is hoping that it will have an effect on the recipient.

These four levels of communication exist on both sides, on the sender's side as well as on the receiver's side. Therefore, the probability is relatively high that what was told by the sender is not necessarily

Figure 20.2. What are the four messages in this communication? © 2015 Jakob K., reproduced with permission.

what was heard by the receiver. Let's take a look at an example, a communication between two participants in a business context (Figure 20.2).

Of course, it depends very much on the tonality in what intensity each of the four levels of the message are perceived by the receiver. The sender can say it in a factual and sober manner, or with a concerned and worried undertone. Furthermore, especially the relationship level and the self-revelation level depend significantly on the nonverbal expressions (gestures, mimics, etc.) of the sender. In contrast, the factual information level is simple and evident in this case (Table 20.1).

By looking at the relationship level we can discover that there seems to be a disturbance on that level between the sender and the receiver. The sender thinks highly about the receiver, but the receiver hears it as a kind of mistrust on that level. Such disturbances on the relationship level can be critical, because the relationship level dominates the factual information level. Paul Watzlawick (1921–2007), who was an Austrian–American psychologist, communications theorist, and radical constructivist, stated that communication has a content and a relation-

TABLE 20.1 The Four Messages in the Communication Depicted in Figure 20.1

	Sender	Receiver
Factual information level	*The deadline for this document is Friday*	*The deadline for this document is Friday*
Relationship level	*You know that I have a high opinion of you and your work.*	*She apparently thinks that I don't stick to our agreements.*
Self-revelation level	*I'm under pressure. I need your help!*	*The customer puts him/her under pressure.*
Appeal level	*You have to have it done until Friday under all circumstances!*	*She wants me to give the highest priority to the document.*

ship aspect such that the latter classifies the former and is therefore a metacommunication [66]. That implies that if the relationship level is disturbed, a communication on the factual level is not possible. Therefore, it is important to talk about the disturbance first, and the information must wait until the relationship between sender and receiver is clarified and recovered if necessary.

Knowing about these four levels in direct communication could be very helpful to avoid miscommunication. You can actively avoid that your conversational partner gets angry, offended, or sad. And you may be able to guess why your counterpart reacts surprisingly differently on your message than expected.

20.1.3   Factors Influencing Communication

20.1.3.1   The Language. Language affects almost all aspects of everyday life. We need our language to express emotions, share feelings, tell stories, and convey complex information and knowledge. This is also true for communication between all individuals involved in a systems engineering project. Because of language barriers, stakeholders may struggle to communicate what they need or even get necessary information regarding the system of interest. And due to globalization leading to worldwide distributed development teams, language is even more important.

Due to the international nature of many projects and the historical roots of the discipline, English is the dominant language in systems engineering. But it's usually not the mother tongue of all involved people.

This could lead to communication problems and misunderstandings. In addition, language is also always part of the culture. We will deepen this issue in Section 20.1.3.4 when we discuss the various connotation of words, and in Section 20.3 when we talk about collaboration skills in an intercultural environment.

20.1.3.2 The media used. Ideally, communication should always take place face-to-face, that is, the sender transmits an information directly to the receiver, typically through language. This is also known as direct communication. But from an economic perspective this would be too time-consuming and costly for an organization. In a work-related context it is plain not feasible to exchange every necessary information via direct communication. Thus, the direct communication is often replaced by an indirect communication, also called media communication, which is characterized by the use of different kinds of media. Examples include phones or e-mail.

The indirect communication usually doesn't have a nonverbal communication level. There is no visual contact, that is, you can't see the facial expressions and the gestures, which are also known as body language, of the other communication participants. Exceptions to this are video conferencing systems (see Section 20.1.4).

Furthermore, the feedback is often delayed in this kind of communication. Thus, it can sometimes take days or weeks to get an answer to a question by e-mail.

20.1.3.3 Spatial Distance. Already in his seminal 1977 book "Managing the Flow of Technology" [6], Professor Dr. Thomas J. Allen, an expert for organizational psychology and management at Massachusetts Institute of Technology (MIT), describes a strong negative correlation between spatial distance and communication frequency. The so-called Allen Curve (see Figure 20.3) is a graphical representation that shows this correlation. It is the result of a research project by Prof. Allen in which he explored how the distance between engineers' offices affects the frequency of technical communication between them.

The Allen curve depicts that we are much more likely communicate with someone who is located in our immediate proximity, for example in the same room, as with people that sit further rooms away, on a different floor, or even in a different building. With other words, long distances between members of an organization are detrimental to communication.

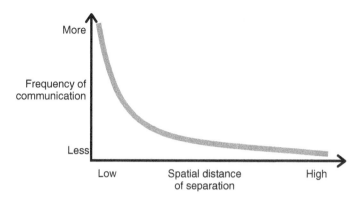

Figure 20.3. The Allen curve depicts the negative correlation between frequency of communication and spatial distance.

Nowadays, we can access a virtual office from everywhere and at any time. Therefore, one might suspect that today's technologies, like e-mail, cloud-based office systems, or video conferencing systems, are able to bridge the spatial distances and the Allen curve is not valid anymore. But that's wrong. Recent researches and studies have shown that the Allen curve is valid not only for direct communication, but also for digital communications. Ben Waber et al. [144] were able to prove that the frequency between colleagues that are working in spatially close distance was four times higher than the frequency with colleagues in distant locations.

In the space industry, it was recognized a few years ago that a close collaboration of engineers is of great importance for the success of a complex mission. Therefore, the European Space Agency (ESA) built a Concurrent Design Facility (CDF) in Noordwijk (Netherland) that allows a team of experts from several disciplines to work concurrently in a highly collaborative environment. The architecture of the building is especially designed to support concurrent working in teams. The CDF consists of several rooms of different sizes and for different purposes. The main design room in the center of the facility, and all other rooms, are equipped with state-of-the-art computer workstations and linked with each other via a high bandwidth network. All design and auxiliary rooms are not only equipped with video conferencing systems but also an audiovisual network allows to display the data from any screen or workstation on any or all of the screens in the other rooms.

20.1.3.4 *Various connotations of words.* Connotation is defined as an implied meaning that is associated with a word in addition to its literal meaning, the latter is also known as denotation. The denotation is neutral and can simply be looked up in an encyclopedia. The implied meaning instead depends upon the social, cultural, and personal experiences of individuals, or it could have emotional reasons. The connotation is an association which the word evokes in a reader's mind, either positive or negative.

As an example, you can say either vagrants or homeless when you talk about people with no fixed address. The first one (vagrant) has a negative connotation and is often associated with a public nuisance. Whereas when you say homeless, this often evokes associations with a need for help, pity, and charity.

The Animal Farm, an allegorical and dystopian novella by English novelist George Orwell (1903–1950), contains many connotations. Many species of animals in the story have a connotation. For example, the pigs connote powerful, corrupt people. "Boxer," a loyal, kind, dedicated, hard working, and respectable workhorse, connotes the laborer class.

Words and terms may have different connotations in different cultures. This circumstance often leads to unintended, sometimes embarrassing misunderstandings. A noticeable example are colors, which have different meanings in different cultures. For instance, the word white, denoting a color, is associated in the Western culture with pure, noble, moral goodness, and innocence. In China, by contrast, the color white stands for pale, weak, and without vitality. Therefore, the bride of a wedding wears a white dress in Western cultures, but in China a red dress.

Furthermore, over time the connotation of words can shift. For example, the word "drug" was originally a neutral term for various active substances, including those that were prescribed by doctors. The term "drug store" was coined in those times. Nowadays, the word "drug" is usually associated with dangerous addictive substances. Hence, in a medical context the words "medicine" or "medication" has replaced "drug".

20.1.4 The usage of communication aids and tools

In addition to sheer personal conversations there are many possibilities to facilitate and promote communication within an organization, or tools to support the direct communication. Systems engineers can take great advantages of these possibilities if they are used in the right way.

- *E-mail:* During the last decades, communication via electronic mail (short: e-mail) has not only largely replaced ordinary letter mail. It can also be seen as a generally accepted alternative to other types of communication, like direct communication or telephone calls.

  Nowadays, sometimes e-mail as a media is even considered harmful. The huge amount of e-mails that are received by many employees every day could be overwhelming. Several hundred e-mails a day are no rarity. The processing of the messages is time-consuming and annoying. Studies have shown that the permanent sending and receiving of e-mails is setting employees under stress. Some companies have even announced plans to waive e-mail in the medium term.

  E-mail should not be used inconsiderately. Before sending an e-mail you should consider carefully whether it is really necessary. Especially, functions like "reply to all" or "carbon copy" (cc) can cause unpleasant situations. The message may be, for example, sent to not intended addressees. Furthermore, a more economical use of these functions prevents an e-mail flood.

- *Blogs* and *Wikis:* During the last years, weblogs (short: blogs) and wikis[1] became an important part of knowledge sharing in organizations. Both applications are usually web-based and accessible for members of an organization through the organization's intranet. What makes both applications so powerful for collaboration and communication is its "anyone can edit" capability.

  Due to their chronological nature, blogs are usually used like a diary. For instance, a project blog can be used by members of a systems engineering project to regularly inform team members and other stakeholders about project progress, important events, new insights, relevant issues, and so on. This creates transparency in the project, but the information are usually only accessible for a limited set of users within the organization.

  In contrast to this, wikis behave more like a reference book. They are more suitable for knowledge management for the whole organization, across departments and team boundaries, and beyond end of projects. In some organizations, the wiki is the single source for all company information.

  Wikis and blogs live from participation. So that individuals can get the information out of these applications, they must actively access them. Problems could arise if stakeholders don't actively access the

---

[1] The term Wiki is derived from the Hawaiian word "Wikiwiki" which means "quick."

content, because this would lead to a standstill of the information flow via these channels. Therefore, both blogs and wikis allow notification of changes to be sent out via RSS feed.[2] This lets users interested in a specific topic to be notified when changes are made in the wiki or blog. And it could also ensure that the interest about the contents remains high.

- *Instant Messaging (Chat):* Instant Messaging (IM) provides real-time text transmission over the Inter- or Intranet. It is very similar to Short Message Service (SMS), a service that allows fixed line or mobile phone devices to exchange short text messages.

  The times when Instant Messaging was solely used for personal chat and entertainment are long gone. The use of IM within the business environment is growing at an exponential rate. It provides a powerful possibility to stay in contact with employees, customers, and vendors.

  Like any kind of communication via the Internet, also the use of Instant Messaging is not without risk. In addition to several security aspects (viruses, worms, malware, intentional or accidental revelation of confidential material or intellectual property, etc.), it is also not suitable for every communication. It is often more intimate than other electronic media like e-mail, and the communication style is often characterized by Internet slang and shortcuts. This does not always fit to a business context, and it can also lead to misunderstandings or sensitivities.

- *Telephone-/Videoconferencing:* In times of globalization and worldwide distributed development teams, telephone- and video-conferences are broadly accepted as an alternative to direct communication and can significantly reduce the need to travel. Since the availability of high-capacity broadband telecommunication services at relatively low costs, video conferencing has undergone a widespread in many domains, like business, education, engineering, medicine, and media. The range of different video conferencing systems on the market is high. In addition, some social networks also provide video conferencing features.

  One of the strengths of many of these systems is the possibility to share the computer desktop or an application with all participants. This allows, for instance, collaborative work with a system's architecture model.

---

[2] RSS is an abbreviation for Really Simple Syndication and describes a file format that allows to publish frequently updated information.

- *Information radiator:* The term information radiator was first coined by Alistair Cockburn around the year 2000 and is described in his book "Crystal Clear" [22]. This kind of communication aid is very well known in the agile movement. Cockburn describes an information radiator as a well-readable display that is posted in a prominent place where people can see it while they work or walk by. A good information radiator is large, good readable, and can be conceived by observers at a glance. Usually, it is either a board, a poster, or sometimes a screen. It is a kind of one-way-communication and typically used to show status information.

  An example for an electronic information radiator can be found in many software development projects. There, a flat screen on the wall displays a dashboard that visualizes the statuses of the automated continuous build and integration processes.

- *White Boards* and *Flipcharts:* Especially in workshop situations, whiteboards and flipcharts are excellent tools to work together in a highly collaborative way. For instance, system architecture drafts can be sketched and discussed. Questions can be clarified immediately, and all participants could develop a common understanding about the system's architecture.

  The whiteboards or flip charts can be photographed and archived after the session, and the images can be embedded in a blog or wiki article. Alternatively, the sketches can be transferred into a SysML modeling tool.

  In Section 14.9, we show the usage of these tools for the collaborative development of a first functional architecture for a system within a workshop.

20.2 PERSONALITY TYPES

All humans are equal, but yet very differently. Since the dawn of time, humans have tried to describe and categorize their personality in many ways. This in turn led to a variety of models, all of them trying to fit something complex and fluid as the human's personality into certain categories. As already noted in the introduction to this chapter above, all of these models are more or less wrong. No matter which personality types model you use, no one on this world is either exactly "type A" or "type B." Like all models, they represent a simplification which in no way is able to represent the complex reality.

Furthermore, it is important to bear in mind that models about personality types are just an aid to predict vaguely how people are likely to behave in certain situations. Those models are inappropriate to determine accurately how a person will behave in any case and under all circumstances. And they should not to be taken as gospel.

20.2.1 Psychological types by C. G. Jung

One famous model dates back to early twentieth century and was the brainchild of Carl Gustav Jung (1875–1961), often referred to as C. G. Jung, who was a Swiss psychiatrist and psychotherapist. C. G. Jung's model of psychological types was first published 1921 in "Psychologische Typen" [68] (engl.: "Psychological types" [69]) and is perhaps the most influential theory in personality typology.

According to C. G. Jung's theory, people can first be characterized by their preference of general attitude. These preferences are **Extraverted (E)** versus **Introverted (I)**. In psychology, an extraverted person is more concerned with practical realities than with inner thoughts and feelings.

In addition, people can be characterized by their preference of one of the two functions of perception. These are **Sensation (S)** versus **Intuition (N)**. Sensation means that a person pays more attention to information that comes in through his/her five senses. In contrast, intuition means that a person pays more attention to the patterns and possibilities that she sees in the received information.

The third pair describes the preference to one of the two functions of judging, that is, it characterizes how a person like to make decisions. These are **Thinking (T)** versus **Feeling (F)**. A thinking person puts more attention on objective principles and impersonal facts. A feeling person likes to do whatever will establish or maintain harmony.

The preferences E and I, the perceiving functions S versus N, and the judging functions T versus F were combined by C. G. Jung to eight psychological types as given in Table 20.2.

C. G. Jung stated that we define our psychological type by our so-called *dominant function*.

For instance, if we like to use Extraverted Sensation more than the other functions, then we are an Extraverted Sensation type. We like to gather factual data then, and use our senses to see, feel, touch, smell, and listen what's going on in the world. People who are Extraverted Sensors process life through their experiences and are able to live in the here and now.

TABLE 20.2   The Eight Psychological Types by C. G. Jung

	Perception	Judging
Objective	Extraverted Sensation	Extraverted Thinking
Subjective	Introverted Sensation Extraverted Intuition	Introverted Thinking Extraverted Feeling
	Introverted Intuition	Introverted Feeling

In contrast, if our dominant function is Introverted Feeling, then we like to make decisions based on emotions rather than on objective facts and data. People who are Introverted Feelers have the ability "to see through others" and the highest level of empathy of any types.

The American psychologist Katharine C. Briggs (1875–1968) and her daughter Isabel Briggs Myers (1897–1980) took up C. G. Jung's work and developed his theories further. The result of their research is an expanded model with 16 distinctive personality types, and an assessment method named Myers–Briggs Type Indicator® (MBTI®).[3] MBTI is an instrument for people to discover their personality type. A detailed discussion of MBTI is beyond the scope of this book, you can find more information on the website of the Myers & Briggs Foundation [134].

The practical application of this knowledge is diverse. Knowing about psychological types can be useful for career planning, communication, education, coaching, and counseling. It can improve our awareness of understanding of reactions of other people to situations.

20.2.2   The 4MAT System by Bernice McCarthy

In the role of a system architect, you are frequently in the situation to explain stakeholders the system's architecture and to justify your architectural decisions. This is usually done in a presentation. For your audience is your presentation in most cases a learning situation, as they have to learn and understand new things. Individuals differ in how they learn, they have different learning styles. This fact has been recognized and scientifically studied by the American educational theorist David A. Kolb PhD in the 1970s. Kolb developed a model about different learning styles and published it in 1984 [77]. The model served as the foundation

---

[3] Myers–Briggs Type Indicator and MBTI are trademarks or registered trademarks of The Myers & Briggs Foundation in the United States and other countries.

for Kolb's experiential learning Theory (ELT). Kolb is renowned in educational circles for his assessment tool named Learning Style Inventory (LSI).

Based on David Kolb's learning styles model the American educational scientist Dr. Bernice McCarthy developed the 4MAT System [95]. McCarthy has figured out that people ask themselves at least one of the following four basic questions while they are learning:

- *Why? (motivation/the philosopher)* – Typical questions: Why should I learn this? Is it beneficial for me?

  This type of learner scrutinizes the sense and wants to know why your presentation about system architecture should be of interest to him/her.

- *What? (knowledge/the scientist)* – Typical questions: What are the facts? Can I have more details, please?

  This type of learner wants facts, data, and would like to get an explanation of the thing (e.g., properties of the system's architecture).

- *How? (demonstration/the practitioner)* – Typical questions: How does it work? Can you please show it to me?

  This type of learner wants to get deep knowledge about how something works, and would like to try immediately everything out (e.g., build and test the system according to your architecture).

- *So what? (debriefing/the visionary)* – Typical questions: Where can I apply it? What if … ? Is it possible to … ?

  This type of learner seeks for more, explores hidden possibilities, and thinks about future scenarios (e.g., extensions of the system's architecture)

Knowing about these four types can be a great advantage while preparing and performing your presentation. With this knowledge, you can structure your presentation so that you can satisfy the needs of all four types of learners. You should start with a Why-part to get those ones of your audience who demand applications and examples. In the next part you should provide many facts about your architecture for What-people. Then you should provide details about how it works for the How-people, for instance, you can start a simulation in a modeling tool that demonstrates how the system will work. Finally, you should provide an outlook for the So-What-people, such as future extension possibilities of the architecture.

A further advantage of this knowledge is that we can avoid misunderstandings. If communication problems occur, I can try to classify my counterpart in one of the four categories and adjust myself accordingly. Otherwise, it could happen that What-people and How-people can come into conflict.

20.3 INTERCULTURAL COLLABORATION SKILLS

Today's organizations are often multicultural networks. While large organizations are spread across different continents and thus automatically have local sites near the roots of different cultures, small organizations will still need to do a worldwide search for experts to hire or suppliers to subcontract. We saw in Chapter 10 that system architects communicate and collaborate with multiple stakeholders in the organization's network. Today this often involves meeting different cultures. Therefore, intercultural collaboration skills are needed. In this section, we will therefore briefly discuss intercultural collaboration. The bad news is as follows: even Hampden-Turner and Trompenaars, leading researchers in investigating multicultural collaboration, have to admit about today's two major research points of view that they are "inadequate" [50]. They encourage their readers to think for themselves. So this is what we do as readers of Hampden-Turner's and Trompenaar's publications. Based on our own observations, conclusions, and experience, we present our own pieces of advice for those entering multicultural dialog:

- *Expect "culture" to have multiple dimensions:* So far we have talked about multinational networks, which are the most trivial example of a multicultural environment. However, "culture" is not only something we can attach to the provenience of individuals in a geographical or national sense. We can also see different corporate cultures, different cultures in different political parties, different cultures in different religions and even different cultures in software engineering education as opposed to mechanical engineering education. All these kinds of cultural differences can lead to misunderstandings that in our opinion often have their root cause in missing awareness about the difference. It is this awareness that can help overcome misunderstandings, as we see in the next paragraph. A very rough advice can be as follows: In the first place, do not expect anyone to have the same cultural background as you have, even not people who share your mother tongue.

- *Identify common goals:* Once I am aware that others have a different approach, I can maybe see that in some areas I am striving for the same goal as they are, just in a very different fashion. Once I am able to identify common goals with my dialog partners, it becomes much easier to negotiate those topics around which the alignment of goals is still pending.

- *Expect others to be different:* We have seen many situations in which people from culture A met people from culture B and just went into normal business as they would do it with a member of the family that has been their family's neighbor for the last three generations in the village in which they grew up. In some of these cases, we later heard about difficulties in reaching consensus. If you have reason to believe that people with whom you enter dialog have a different cultural background than you, better be prepared for the extra need for clarifications and be extra cautious to stay open-minded and flexible in the way the dialog is conducted. If you do not understand what "they" say, just ask what they mean.

- *Avoid using metaphoric language in multicultural dialog:* As discussed earlier in Section 20.1.3.4, different people can associate different meanings to the same word. This is even more relevant for metaphors, because they often make an implicit reference to the literature or the history of one culture. For example, the phrase "to meet one's Waterloo" means to meet a major defeat. It makes implicit reference to the final defeat of a person called Napoleon that was the result of a battle around a place called "Waterloo" [76], which today belongs to Belgium. The British named "Waterloo Station" according to this event in which they were on the winner side. Many British can probably relate to the phrase. Napoleon was French. Since none of the authors has this nationality, we do not even dare to predict how a French man or woman would feel about the words "to meet one's Waterloo" - and in most of the world these words would probably have no meaning at all. We recommend to avoid any metaphor in multicultural dialog. Since metaphors are supposed to explain something, it must be possible to explain the same thought in simple words and explicit technical terms instead of obfuscating it with a metaphor. Instead of talking of "meeting one's Waterloo," we can talk of "failing."

- *Be careful with humor or irony:* Trompenaars and Hampden-Turner [137] discourage the use of irony, because it is based on saying the opposite of what is meant. They also report different perceptions of humor by business people from different cultures. Even though humor can be a good facilitator in a well-established team with a mutual understanding of the limits of what is funny, we thus recommend to be very restrictive on the use of humor in new intercultural setups.

- *If you think "they are stupid" then you may be trapped in ignorance:*After meetings between people from different cultures, we may interpret from some participants' evaluation that they considered the other participants to be unqualified or not applying the right mindset to the task at hand. Often, it indeed occurs that approaches between different meeting participants are *different*, but this does not automatically imply that one approach is less appropriate than another one. On the contrary, being able to find different approaches is a strength, because it widens the solution space. We have seen differences in the perception of what is the most intuitive approach come up when cultures collide, and we recommend to everyone who thinks "the others' approach is stupid" to rethink into "their approach is different." This is a first step for getting into effective collaboration with those proposing the different approach. If you find yourself thinking that others propose a stupid approach then ask them why they suggest the approach rather than to finalize your conclusion that you will not pursue it.

- *Read Kissinger:*We already referenced the book "Diplomacy" by Henry Kissinger [76] when talking about Napoleon's final defeat. Even though it is about multicultural phenomena in world history and politics, it gives some insights on typical multicultural misunderstandings that should be avoided. You may perceive that Kissinger expresses his very own point of view in his book. In case this happens and you disagree with that point of view, then this trains you to accept the existence of different points of view, which is a nonnegotiable prerequisite to enter any dialog, especially the multicultural one.

Last but not the least, keep remembering that we need multiple cultures and diversity to succeed with our business. Probably intercultural dialog will still stay one of the challenges in future and will not always be easy, but as long as we keep assuming that multicultural work produces more competitive results than monoculture, we can hopefully motivate ourselves to enter the difficult but necessary dialog with other cultures.

Chapter 21

Outlook: The World after Product Line Engineering

So far, each methodology has its time to come and its time to be replaced by another methodology. At the moment, we see product line engineering emerge in different organizations, and we believe that these organizations achieve return of investment from architecting a whole product line at once instead of a single product. We are convinced that research around product line engineering will provide even more efficient product line approaches than we know today.

Yet one should be aware that each approach will be obsolete at a certain point of time in the future. This will be the case with product line engineering sooner or later. What comes next? Of course we do not know, but we will draw a scenario here that should start a debate about the advantages and disadvantages of product line engineering.

Have you ever seen museum robots from manufacturers producing product lines? Have you been in a situation in which you thought you noticed that the museum robots of different brands from the same organization look similar? Have you seen expensive museum robots that are almost the same as their cheaper counterparts, just a little bigger and with more expensive materials used on easily exchangeable parts? We do not know. In case you have seen such museum robots

Model-Based System Architecture, First Edition.
Tim Weilkiens, Jesko G. Lamm, Stephan Roth, and Markus Walker.
© 2016 John Wiley & Sons, Inc. Published 2016 by John Wiley & Sons, Inc.

or can imagine them, ask yourself: Will the customer feel comfortable with this or fooled?

Imagine a company that disrupts the whole market by investing all development efforts in tailoring a product to exactly one market segment. No reuse of modules and no generic architecture is encouraged, only the achievement of optimum fit to the market segment. If the company is successful in doing this and in selling the resulting product, it has at least the chance to gain high market share worldwide. Let's call this product a full-custom product.

Table 21.1 compares this hypothetic scenario with a product line scenario that could be applied to serve the same market segment. The

TABLE 21.1 Product line engineering versus gaining market leadership with a full-custom product

Criterion	Reasoning in Favor of a Product Line	Reasoning in Favor of a Full-custom product
Economies of scale	Economies of scale are achieved by reusing parts and subassemblies across different market segments.	Economies of scale are achieved by becoming worldwide market leader and selling a high number of units.
Reaction to changed market needs	By change management on the product line scope, the high cost of adapting is managed carefully.	New market needs can be handled in an agile way within the scope of one product, without having to analyze the impact on the whole product line.
Addressing customer needs	The degree of fulfilling customer needs is traded against variability cost.	The degree of fulfilling customer needs is traded against development cost.
Optimizing the product	Optimization for minimum deviation from needs across market segments.	Optimization for the given market segment.
Customer satisfaction	Customers have the option to choose among a variety of products.	Customers have the option to buy the best product in the market segment.
Reuse of parts and subassemblies	Reuse is architected into the product line.	Reuse can be established later, when full-custom products are successful in different market segments and yet expose some commonalities.

whole comparison is of course highly hypothetical and neglects most aspects that one would need to consider to make a realistic business case out of the full-custom product. We are not showing this table to make anyone believe in easily becoming world market leader, we are rather making it to inspire you dear reader to challenge the current systems architecting approaches like product line engineering and all the other nice concepts we have presented in this book. Choose them thoughtfully and with a good reason.

We do not know if the research in product line engineering will keep up with the individual needs of different market segments and markets. We also do not know whether there will be a need for individualized mass products in future, or whether future generations of buyers will ask for massive replication of exactly the same product configuration in order to obtain cheap technology for all.

All these factors determine how the match-deciding ball game in systems architecting will look like in future. What we do believe is that the systems architecting work whatever it will be like is going to be one of the key factors for selling better products successfully - hopefully based on model-based systems architecting.

Appendix A

OMG SysML

The OMG Systems Modeling Language (OMG SysML) [105] is a modeling language for model-based systems engineering. It supports and enables the specification, analysis, architecture and design, and verification and validation of a system. SysML defines the notation, semantic and abstract syntax (data structure) of the model elements, and a set of diagrams as views on the model. The diagrams are clustered in structure and behavior diagrams, and the requirements diagram (Figure A.1).

SysML is based on the Unified Modeling Language [106]. Both languages are defined and maintained by the Object Management Group (OMG). In 2001, the International Council on Systems Engineering (INCOSE) decided to establish UML as a standard modeling language for systems engineering. At that time, no standard modeling language for systems engineering was available and UML was already widely spread and used in software engineering and partly in systems engineering. Tools, educated engineers, and best practices for UML were available. To avoid cluttering the language, they decided against adding a systems engineering perspective to UML. Instead, a new modeling language using the profile extension mechanism of the UML should be developed. As a result of the standardization process SysML 1.0 was published in 2006 as a OMG standard [105]. Formally, SysML is a

Model-Based System Architecture, First Edition.
Tim Weilkiens, Jesko G. Lamm, Stephan Roth, and Markus Walker.
© 2016 John Wiley & Sons, Inc. Published 2016 by John Wiley & Sons, Inc.

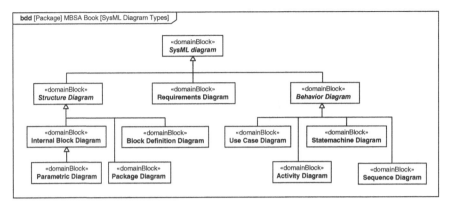

Figure A.1. SysML diagram types.

profile of UML, that is, a special extension of UML (Section A.5), although it is treated like a modeling language on its own.

SysML adds new model elements that are missing in UML like requirements or that are specific for systems engineering like blocks. SysML also removes elements from the UML vocabulary that are not useful in systems engineering like classes or components that are specific for software engineering.

In summary, SysML is much smaller than UML, for example, SysML has 9 and UML 14 different diagram types. In the following, we give a brief description of each SysML diagram type. A detailed description of SysML including a modeling methodology can be found in Ref. [145, 147]. This chapter gives only a brief overview and cannot replace a complete book or training about SysML.

A.1 DIAGRAM AND MODEL

SysML follows the principle of separation between the model and the representation of the model, also called "View and Model" (Section 7.7). The language defines a semantic and abstract syntax (model) and a notation or concrete syntax as well as a set of diagram types (view). The information of a SysML diagram like position or size of an element is stored separately from the model information. The left side in Figure A.2 depicts the abstract syntax. For instance the use case "Book a Tour." The concrete syntax, that is, the ellipse notation of a use case, is not shown. The right side in Figure A.2 shows parts of the appropriate diagram syntax, that is, the standardized data structure of the layout information.

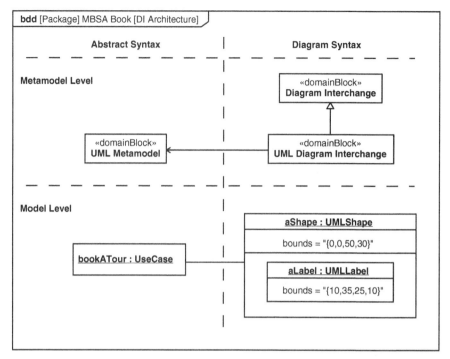

Figure A.2. Diagram interchange.

Views show only the information that is intended by the model builder. They are incomplete according to the information that is stored in the model. The separation of the view and model is obvious for example in an application like Excel where the data in the sheets is the model and the diagrams are the views on the data. The author edits the information directly in the model and the diagrams are exclusively used for presentation purposes.

It is different for SysML. Typically, the model builder creates and edits information using a diagram and not the model itself. The model gets into the background and the primary artifacts are the diagram and its elements. Therefore, it is important to emphasize that a SysML diagram is only a view and the real information lies in the model.

The nine SysML diagram types are standard view types on the model. However, it is not forbidden and explicitly desired that you create additional views, for example, PDF documents or spreadsheets.

The header of a SysML diagram follows the syntax `<diagram type> [ <model element> ] <model element name> [<diagram name>]`. In Figure A.2, the diagram type is "bdd" (= block definition diagram), the designated model element is a package, that is, the diagram shows the content of a package, the name of the package is "MBSA Book" and the name of the diagram is "DI Architecture."

A.2 STRUCTURE DIAGRAMS

The structure diagrams are a set of diagrams to provide a view on the structure of the system (Figure A.3). The main diagrams are the block diagrams that present the structural aspects of a system. Block diagrams are the block definition and the internal block diagram.

The package diagram shows the namespace structure of the model. The parametric diagram is a special internal block diagram and shows the parametric relationships between value properties of the system.

A.2.1 Block Definition Diagram

As stated in the name, the block definition diagram depicts the definition of blocks, that is, the blueprints of the physical or virtual system parts. The main element is the block itself that defines the structure and behavior of an entity. A block has value properties, operations, constraints, and owned or shared properties defined by other blocks.

Figure A.4 depicts the block "Museum Robot" with value properties *range* and *mass* including units, a constraint for the weight and an operation "getCapacity()" that returns the battery capacity of the museum

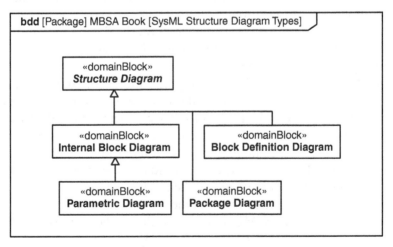

Figure A.3. SysML structure diagram types.

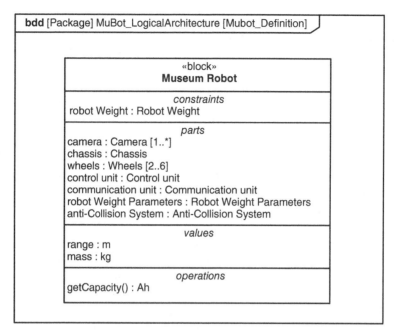

Figure A.4. SysML block.

robot in Ah. The parts of the block are typically defined with a composition relationship in the form of a directed association with an arrow at
the end of the defining block of the part and a black diamond symbol at
the end of the owner of the part. Figure A.4 shows the parts in a special
compartment of the owning block. Figure A.5 shows the parts of the
Museum Robot with the composite relationship notation.

Besides the association, the generalization is another common
relationship in a block definition diagram. It is a solid line with a hollow
triangle as an arrowhead. You see a generalization relationship for
instance in Figure A.5 in the upper left corner between the "Virtual
Tour Server" and the "VMT Server." The Virtual Tour Server is an
element of the logical architecture and specializes the abstract VMT
Server from the base architecture. A generalization in one direction is
a specialization in the other direction. It depends on your viewpoint.
An abstract block means that the definition of the block is too generic
and it needs additional information for a concrete definition of the
entity. It is depicted by a italic font of the name. The Virtual Tour
Server inherits all features from the abstract VMT Server due to the
generalization relationship. Simply spoken, all features are copied
from the general block to the special block. It is possible to redefine

inherited features. In summary, that means both blocks together are the complete definition of a Virtual Tour Server entity.

Two common applications of the block definition diagram are the product tree and the domain knowledge model. The product tree is a tree-structured breakdown of the system (Figure A.5). The diagram focuses on the list and hierarchy of the system blocks. Therefore, the details of each block are typically hidden.

The domain model defines the domain knowledge of the system (Section 8.5). Figure A.6 shows an extract of the domain model. A more complete diagram is shown in Figure 8.9. The blocks of the domain model have the stereotype «domainBlock» from the SYSMOD profile [145, 147] to mark them as a special kind of a block. The solid line between the blocks is an association that defines properties at both ends. The respective property relates to the block on the other side of the association. The associations are read in the following pattern: "Every <block> has <multiplicity> <block> in the role <role name> at any time." This sentence must be meaningful for the domain. Let's test it for our example in Figure A.6: "Every customer has any number of tours in the role booked tours at any time." Though this sounds a bit clumsy it makes sense.

The association between the block "Customer" and the block "Tour" in Figure A.6 defines a property named "visitor" of type "Customer" with multiplicity [0..*] and a property named "bookedTour" of type

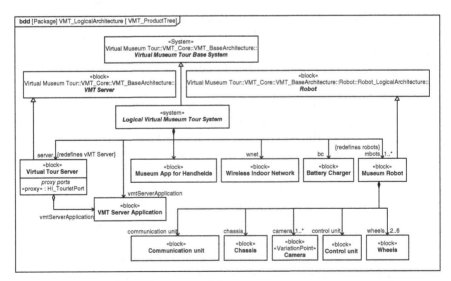

Figure A.5. Example product tree of the VMT.

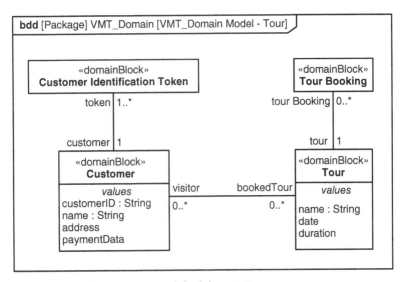

Figure A.6. Example domain model of the VMT.

"Tour" with multiplicity [0..*]. The property visitor is a reference property of the block Tour and the property bookedTour is a reference property of the block Customer. Unlike the composite part property described above, the reference property is not owned by the block.

A special application of the block definition diagram is the activity tree. The whole activity without its internal structure (actions, flows, control nodes) could be shown like a block in a block definition diagram. It is depicted as a rectangle with the name of the activity and the keyword «activity». The activities are modeled in a tree structure with the composition relationship (Figure A.7). The tree depicts a call hierarchy and not a ownership hierarchy like the product tree.

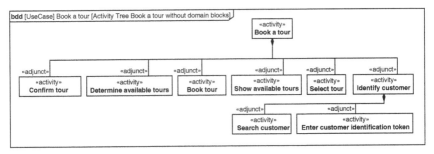

Figure A.7. Activity tree.

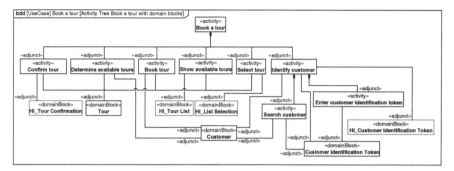

Figure A.8. Activity tree with associated blocks.

The semantic of the composite relationship specifies that the activity at the black diamond side calls the other activity by a call behavior action. It owns the execution instance of it. The properties at the association ends typed by an activity are SysML adjunct properties («adjunct»). They constrain the values of the properties to be determined from the appropriate call behavior action. The input and output parameters of the activity could be depicted with the association relationship as shown in Figure A.8.

A.2.2 Internal Block Diagram

While the block definition diagram depicts the definition of the blocks, the internal block diagram shows the internal structure of a block, that is, the properties, their connections, and the interfaces between them.

Figure A.9 shows a part of the logical architecture of the VMT in an internal block diagram. The rectangles in the diagram are properties

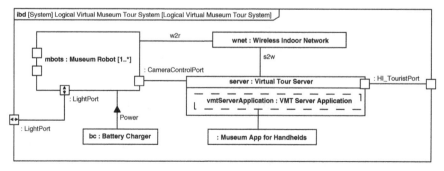

Figure A.9. Example internal block diagram of the VMT.

of the corresponding block. Here the enclosing block is the "Logical Virtual Museum Tour System," that is, the root node of the logical architecture, as you can see in the header of the diagram. The syntax of the header is explained in Section A.1. The diagram frame is the border of the enclosing block.

The properties of the block are also called parts when they are typed by a block and defined as a composite property. For instance, "mbots" is a part of the "Logical Virtual Museum Tour System." A part has a name ("mbots"), a type ("Museum Robot") and a multiplicity ([1..*]). The textual syntax is "mbots:Museum Robots[1..*]." The definition of the properties are shown in the block definition diagram in Figure A.5. Noncomposite properties—so-called reference properties—are shown with a dashed rectangle.

The solid line between the properties is a connector. It represents that there is some kind of interaction or exchange of items between the properties. Connectors could have a name. For instance, "w2r" and "s2w" in Figure A.9.

The exchanged items could be explicitly specified by the item flow. It is depicted by a black triangle to indicate the direction and a text nearby that describes the item. The item is a property of the surrounding block or simply the type that flows. In Figure A.9, the item flow specifies that power flows from a battery charger to the museum robots.

The small rectangles on the border of the property rectangles are ports. They are special properties and are defined at the corresponding block of the property and represents an interaction point of the block to its environment, that is, it is an interface. Like a property a port has a name and a block as its type. The block defines the properties that flows through the port and provided and requested structure and operations. Flow properties are specified with a direction ("in," "out," and "inout") and features like operations are specified if they are provided ("prov") or requested ("reqd") (Figure A.10).

SysML differentiates between full ports and proxy ports. A full port is similar to a part that is placed on the border of the enclosing block. It represents a real element from the bill of material (BOM) that is an interface of the enclosing block. A full port is shown with the keyword «full».

The proxy port is a placeholder that represents internal parts at the border of the enclosing block. It is only a placeholder and not an element of the BOM. The semantic of a full port can also be modeled with a proxy port and an internal part (Figure A.11).

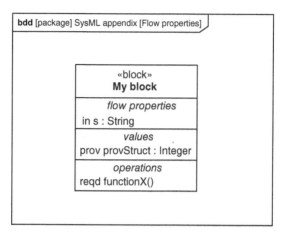

Figure A.10. Flow properties and provided/requested features.

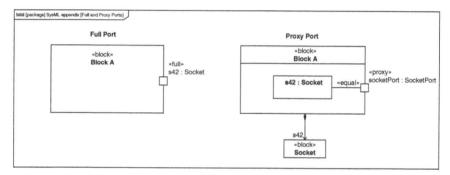

Figure A.11. Full and proxy port.

The connector between the proxy port and the internal part is a binding connector. It defines that the instances of the connected properties are identical, that is, the placeholder proxy port always represents the real element. In practice, the keyword «equal» at the binding connector is often omitted.

We recommend to only use proxy ports and no full ports. You can replace every full port with a proxy port and the appropriate internal part as shown in Figure A.11. But you typically will always need proxy ports in your model to show deep-nested interfaces at the border of a block. Figure A.9 shows proxy ports at the border of our system block "Logical Virtual Tour System" that represents interfaces of inner parts. The system element itself has no real interfaces. If you do not use any full ports in your model, you can discard the keyword «proxy» at the proxy

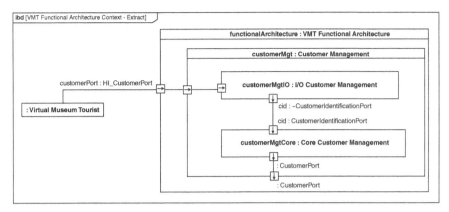

Figure A.12. Behavioral proxy ports in the functional architecture.

port in the diagram. It is no longer ambiguous. Note that the information that it is a proxy port is still in the model. Just the proxy keyword is removed from the diagram.

Instead of being connected to internal properties, the proxy port could also be a so-called behavioral port. That means that any requests arriving at the port are handled by the internal behavior of the block that owns the port. The ports at the inner functional parts in Figure A.12 are behavioral proxy ports. A behavioral port could have an additional small state symbol inside the enclosing block attached with a solid line to the port. We do not use that notation in Figure A.12.

The port *cid* at the part "customerMgtIO" in Figure A.12 is a conjugated port as depicted by the tilde symbol (~) before the port type. A conjugated port reverses the directions of the features and flow properties of the port type, that is, "in" turns to "out," "prov" to "reqd," and so on.

Figure A.13 shows some types of the proxy ports. They are interface blocks that specify provided and required operations and properties as well as flow properties. They are only specifications of interfaces and do not represent real entities. The user interface is a special interface block. «userInterface» is a stereotype of the SYSMOD profile to mark human interfaces that specifies human–machine interactions [145, 147]. To assure that the owning block of behavioral ports implements the behavior, the block is a specialization of the appropriate interface blocks (Figure A.13). The stereotype «conjugated» at the generalization relationship is not part of SysML. SysML does not provide the generalization of conjugated types. The conjugated generalization reverses all direction of the inherited features.

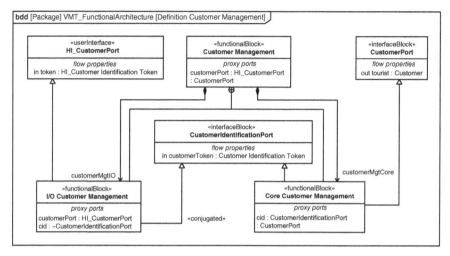

Figure A.13. Interface block specialization.

A.2.3   Parametric Diagram

The parametric diagram depicts relationships between property values of the system. The properties must not be defined in the same block. However, the parametric diagram is a special internal block diagram and the appropriate enclosing block must have access to the properties. Typically, it is the first common node of the properties in the product tree or an extra created context block that represents the context of the constraints.

In Figure A.14, the enclosing block is the VMT context element. It is the lowest node in the tree that has access to properties of the museum robot and to properties of the museum exhibition that is a system actor of our system.

The property values are shown with rectangles like in the internal block diagram. The parametric relationship is a usage of a constraint block that is defined in a block definition diagram. The constraint block defines constraints and their parameters. Figure A.15 shows the constraint block "Range of Museum Robot" that refines the textual requirement "Range." The constraint block has the SysML stereotype name «constraint» above the name. The parameters of the constraints are defined in a special compartment with headline "parameters."

Figure A.15 is a parametric diagram that depicts the usage of the constraint block "Range of Museum Robot." The squares attached at the inside of the constraint rectangle in a parametric diagram represent the parameters of the constraint block. The solid line between a parameter

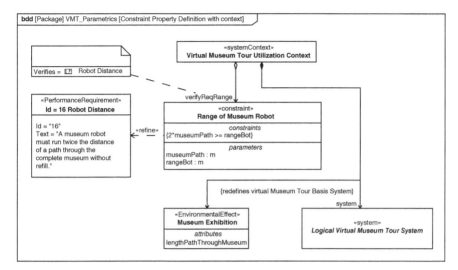

Figure A.14. Definition of a constraint property.

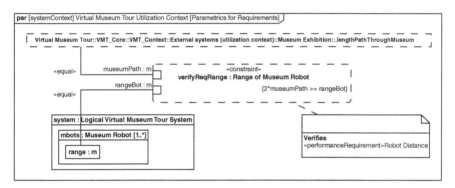

Figure A.15. Example of a parametric diagram.

and a property value is a binding connector that asserts that the values at both ends are equal. Undefined values could be calculated by a modeling tool if the equations are solvable. For example, you can calculate the minimum range of a museum robot by adjusting the parameter for the length of the path through the museum.

The usage "verifyReqRange" of the constraint block in Figure A.15 verifies the requirement in the model. SysML has a verify relationship to model that explicitly. It is a dashed line with an open arrowhead and keyword «verify». Figure A.14 depicts the alternative call-out notation. It can show the information of the verify relationship if only one of the related elements are shown.

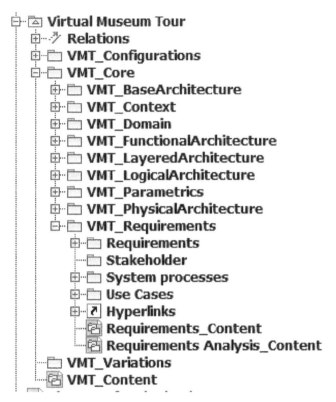

Figure A.16. Model browser of a modeling tool.

A.2.4 Package Diagram

Packages provide a generic model organization capability. Typically, a model consists of hundreds or thousands of elements. They must be structured to get a better overview, to work with several people on a single model, and to reuse parts in other models. Packages are like the directories on your hard disk to organize the files. Similar to file browsers, modeling tools have a model or project browser to show the package structure of the model (Figure A.16). That tree notation in a browser is not part of SysML. Instead, SysML provides the package diagram to show packages and their relationships.

Figure A.17 shows the top-level packages of our VMT model and the application of profiles (Section A.5). The package symbol with the triangle represents the model. It is a SysML model element similar to the package element.

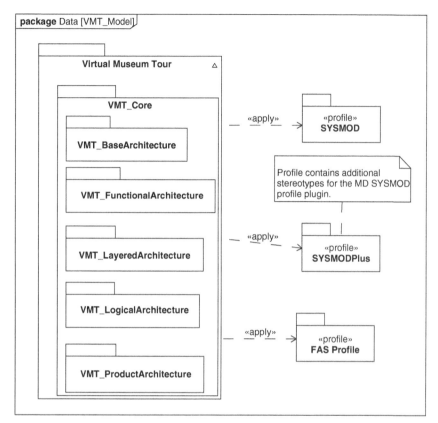

Figure A.17. Example package diagram of the VMT.

Nested packages could be shown inside the owning package (Figure A.17). If you show the nested relationship explicitly you can show the packages in a tree-like notation (Figure A.18).

A.3 BEHAVIOR DIAGRAMS

The behavior diagrams show behavioral aspects of the system (Figure A.19). The activity diagram shows flow-oriented behavior, the state machine diagram event-oriented behavior, and the sequence diagram message-oriented behavior. The use case diagram is an exception. Strictly speaking, it is not a behavior diagram but a structure diagram. It does not show how behavior is specified but a list of top-level system behaviors. Anyway, SysML (and UML) still categorizes the use case

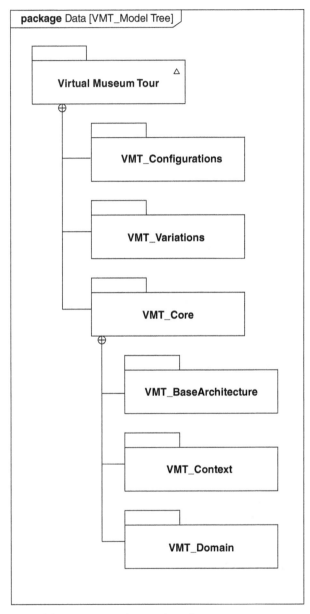

Figure A.18. Tree-like notation of nested packages.

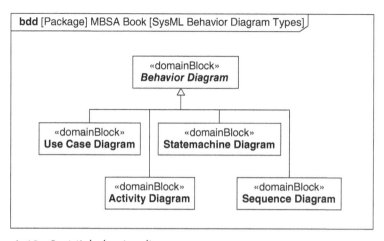

Figure A.19. SysML behavior diagram types.

diagram as a behavior diagram. It is a minor issue and not worth to fix it with regard to the effort and impact of the change in the specification and the modeling world.

A.3.1 Use Case Diagram

A use case specifies a set of actions that leads to an observable result that is of value for the actors or stakeholders of the system. The use case diagram shows the use cases with their associated actors and relationships to other use cases. Typically, the specified behavior of an use case is described by an activity (Section 8.3).

Figure A.20 is a use case diagram that shows some use cases of the actor "Virtual Museum Tourist" of our example system VMT. The sticky man symbols represent actors. It is the standard symbol for actors in SysML. However, it is possible to use other symbols to visualize different actor categories. For instance, a cube symbol is commonly used for nonhuman actors like an external system. The stereotype «user» is part of the SYSMOD profile and not a standard SysML element.

The actors are associated with the use case by an association. It is the solid line that specifies that the actor is involved in the behavior that is described by the use case. It is the same relationship that is used between blocks (Section 298).

«systemUseCase», «secondaryUseCase» and «continuousUseCase» are stereotypes of the SYSMOD profile to categorize use case types and are not part of SysML. See Section 83 for a description how to use these special use case types.

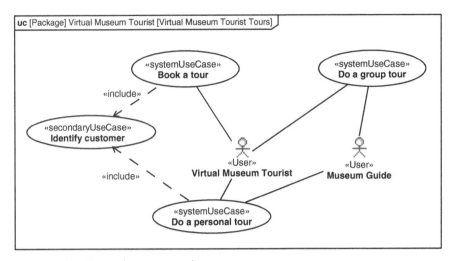

Figure A.20. Example use case diagram.

The include relationship depicted by the dashed arrow with attached keyword «include» in the diagram specifies that the target use case is included in the source use case. That relationship could be used to model use cases that are used in more than one use case to avoid redundancy. That means the behavior of the included use case is also part of the behavior of the including use case.

A.3.2   Activity Diagram

The activity is a special behavior that specifies the execution order and the inputs and outputs of actions. The activity diagram depicts exactly one activity. The order of the actions are specified by control nodes, for instance, fork or decision node, and by control flows. The relationship between inputs and outputs of the action is specified by object flows.

A common application of activities is the specification of a use case behavior. Figure A.21 shows an activity diagram of the use case activity "Book a tour" of the VMT. The two horizontal lanes are activity partitions that group the included actions according to a given criterion. The model builder can define any kind of criteria. Here it is the separation of input/output logic and core logic.

The rectangles at the border of the activity diagram represent the input and output parameters of the activity. The small rectangles at the border of the actions are input and output pins and represent the parameters of the action.

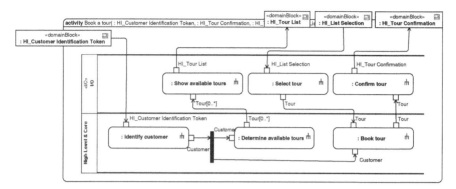

Figure A.21. Example activity diagram.

An action is an atomic piece of behavior. It is not further described in the model. SysML defines several action types in a formal way that makes it possible to provide a common model execution environment for them and to run simulations. To close some gaps in the SysML respectively UML specification, the OMG released the Semantics Of A Foundational Subset For Executable UML Models (FUML) [108] that, besides others, refines the action definition of UML.

If you do not execute your model, only a few action types are relevant. The opaque action specifies a behavior defined in any language. Besides formal languages like a programming language it could also be a natural language like English. Figure A.22 shows an extract of an activity diagram with an opaque action "Identify customer." The specification language is English. It is possible to give the action a name in addition to the specification to avoid programming language code in the diagram and to make it easier for the model reader. In Figure A.22, the name and the specification are identical. Note that we recommend to use call behavior actions. See also Section 8.4.

The actions in Figure A.21 are call behavior actions. They call another behavior. Typically, it is another activity, but could also be a state machine or an opaque behavior. Call behavior actions have a small fork symbol in the lower right corner.

Finally, you sometimes need actions to specify sending or receiving of signals. They are called accept event action and send signal action. Figure A.23 depicts an activity with three AcceptEventActions "Every minute," "Off" and "Sensor data" and a SendSignalAction "Notify." When started, the activity listens for sensor data and off signals and the timer signal "Every minute." AcceptEventActions that receive timer signals have a special notation like an hourglass. Each time when the

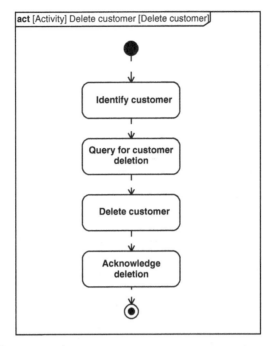

Figure A.22. Opaque actions.

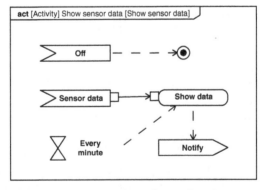

Figure A.23. Accept event action and send signal action.

activity receives sensor data, the data is shown somehow when there is a token at the object flow from the timer signal available. Finally, the activity sends a notify signal. The recipient of the signal is not shown in the diagram but specified in the model. When the activity receives a signal "Off," the activity final node(black circle with outer circle) terminates the execution of the whole activity.

The thick black vertical line in Figure A.21 is a fork node. It is a special control node that splits one ingoing flow into two or more outgoing flows. The subsequent two actions are executed independently of each other and get both the same input object.

SysML provides much more elements for activity modeling. We have just provided the basic set to model most flow behavior of a system. For more information about activity modeling see Refs [145, 147].

A.3.3 State Machine Diagram

The state machine specifies discrete behavior through finite state transitions. It represents the states of an entity and the transitions between the states. The state machine diagram depicts a single state machine.

Figure A.24 shows a state machine of the museum robot from our VMT example. The black dot with the outgoing arrow is the initial state and points to the first state that will be active when the state machine is executed. The first state of a museum robot is "Ready." The state "Ready" has an entry behavior. The action is always performed immediately when the state is entered. It is also possible to define an exit behavior that is performed just before the state is left.

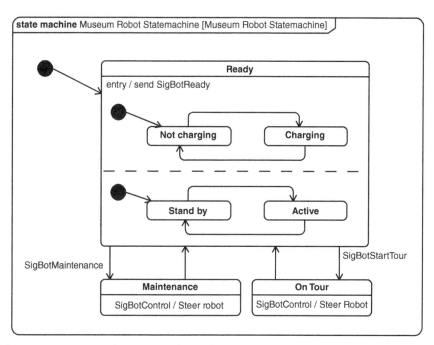

Figure A.24. Example state machine diagram.

If the robot receives the "SigBotStartTour" signal, the appropriate outgoing transition "fires" and the robot is in the state "On Tour." Only in this state and in the "Maintenance" state, the robot reacts on the signal "SigBotControl" with the behavior "Steer robot." It is an internal transition, that is, the corresponding state is not left during the transition. Each state could have substates to model details of the state behavior.

The state "Ready" has several substates in two orthogonal regions. The regions are separated by a dashed line. One region has exactly one active state during execution. For instance, the museum robot could be in the states "Ready::Charging" and "Ready::Stand-by" at the same time. It is not allowed to model a transition from one region to another region. Besides this formal rule, it also makes conceptually no sense to have transitions between regions, because they represent orthogonal concepts.

A.3.4   Sequence Diagram

The sequence diagram shows the exchange of messages between elements of the system. Typically, only one or a few scenarios are shown in a sequence diagram and not all possible paths like in the activity diagram (Section A.3.2), although it is possible to model alternatives, loops, and parallel message exchanges. However, the diagram gets very confusing when you use those elements intensively. Areas of application of the sequence diagram are for instance test scenarios, example scenarios, or detailed specification of a communication protocol between elements.

Like the activity diagram shows an activity model element and the state machine diagram a state machine, the sequence diagram depicts an interaction model element.

Figure A.25 is an example of a sequence diagram of the VMT. It shows a message exchange between the control server and a museum robot. The "Virtual Tour Server" sends the signal "SigBotStartTour" to a robot. The robot responses with the signal "SigBotOk." Then the tour server calls an operation at the robot to steer it to the start position. Again the robot acknowledges the command with the signal "SigbotOk." Finally, the robot responses with the signal "SigBotArrived" when it has arrived at the requested position. In the case of an error, the robot sends a "SigBotError" signal with an error code. The rectangle around the two messages "SigBotArrived" and "SigBotError" is a combined fragment. It is used to model interaction operators like loop

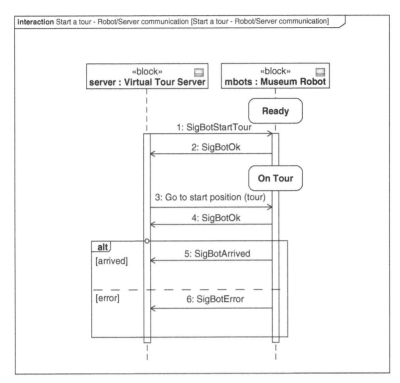

Figure A.25. Example sequence diagram.

or break. Here it is an alternative depicted by the keyword "alt" in the upper left corner of the combined fragment.

The rectangles with the vertical dashed lines are called lifelines. They represent one communication partner. It is an element from the usage level (Section 7.4). For instance a part property of a block. The rectangles on the lifeline represent that the object represented by the appropriate lifeline is active. The arrows represent the messages. An asynchronous message has an open arrowhead as in Figure A.25 and a synchronous message a filled arrowhead.

It is possible to show the respective state of the object by displaying the appropriate state symbols on the lifelines as shown in Figure A.25.

A.4 REQUIREMENTS DIAGRAM

The requirements diagram is neither a structure diagram nor a behavior diagram. It is a diagram category of its own (Figure A.1).

Also in SysML, the requirement is specified by a text. Although you use a graphical modeling language not all texts of a requirements specification are turned into lines and boxes. The requirement text is encapsulated in the SysML model element "Requirement." In addition, the model element stores the name and the unique identifier of the requirement. The applied requirement methodology could add specific requirement attributes like priority or stability. They are not part of the standard SysML and could be added by using the stereotype mechanism (Section A.5).

There are many relationships available to connect the requirement with other requirements or other model elements. For instance, a block could satisfy a requirement, a test case could verify a requirement, a requirement can be decomposed by other requirements, refined by a use case or have trace relationships to other requirements.

The requirement is shown as a rectangle with keyword *«requirement»*. Figure A.26 shows a simple requirements diagram. A requirements diagram is only useful to visualize top-level requirements relationships or to put a focus on very important aspects. SysML provides a table

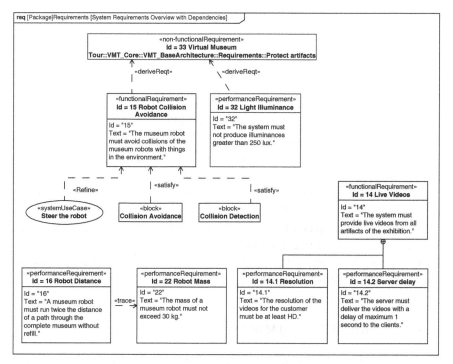

Figure A.26. Example requirements diagram.

Id	Name	Text
2	🖽 Live Videos	The system must provide live videos from all artifacts of the exhibition.
6	🖽 Robot Collision A...	The system must avoid collisions of the museum robots with things in the environment.
7	🖽 Robot Distance	The robot must cover a distance of at least 1 km without recharge.
8	🖽 Robot Speed	The maximum speed of a robot must be 8 km/h.
9	🖽 Resolution	The resolution of the videos for the customer must be at least HD.
10	🖽 Server delay	The server must deliver the videos with a delay of maximum 1 second.

Figure A.27. Example requirements table.

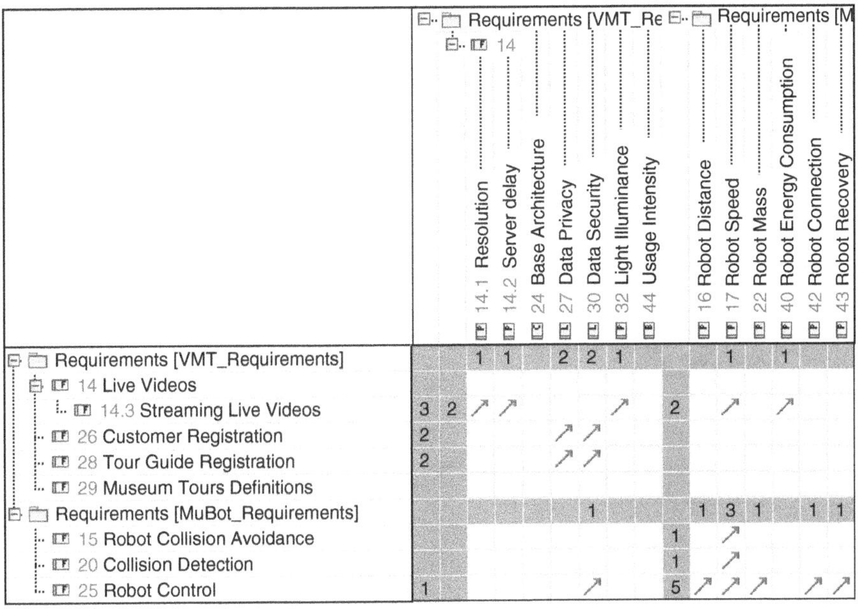

Figure A.28. Example requirements matrix.

representation that is typically used for requirements. Figure A.27 shows a requirement table of the VMT.

SysML also provides a matrix representation to visualize relationships between model element. Figure A.28 shows the relationship between functional and nonfunctional requirements.

A.5 EXTENSION OF SYSML WITH PROFILES

SysML is a common language for any kind of a technical system. Therefore, SysML provides very general model elements that are not specific for a domain, methodology, or system architecture. SysML knows requirements, but not the kinds functional or performance requirements. SysML knows blocks, but no system hierarchy elements like system, subsystem, unit, module, or segment.

In practice, you need a more specialized set of model elements for your specific purpose. SysML provides an extension mechanism to introduce those new model elements to the language. You can't define new elements out of the scratch. They must be based on an existing SysML element and further specify its semantic. Since SysML is also an extension of UML, the base element must be a metaclass of the UML set of model elements. Please refer to special SysML or UML literature for details, for example, Refs [145, 147].

The model element stereotype adds new elements to the model language. It has a name and an extension relationship to the existing UML base element or a generalization relationship to a SysML or other stereotype. The stereotype could define further properties for the model element—sometimes also called tagged values—and a new

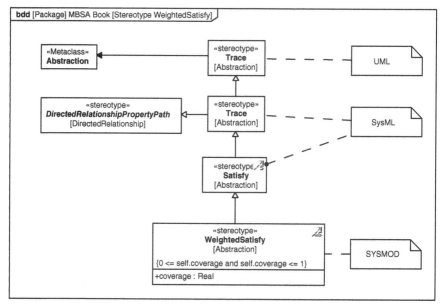

Figure A.29. Stereotype weightedSatisfy.

icon. The semantic of the stereotype, that is, the new model element is described semiformal in natural language or formal with constraints for instance with computable OCL. Figure A.29 shows the stereotype "WeightedSatisfy" from the SYSMOD profile [145, 147]. It specializes the SysML stereotype "Satisfy" and adds a property to specify the coverage of satisfaction. For instance, that a system block satisfies 60% of a requirement and another block satisfies the other 40%. The SysML stereotype "Satisfy" specializes the SysML stereotype "Trace" that specializes the UML stereotype "Trace" that extends the UML metaclass "Abstraction."

A profile is a special package that contains a set of stereotypes. For instance, the SYSMOD profile that contains all stereotypes that are useful for models created with the SYSMOD methodology [145, 147]. The profile is applied with the profile application relationship to a model. That enables the usage of the profile stereotypes in the model. Figure A.17 shows the profile application of SYSMOD and SysML to the VMT model. Note that formally SysML is a profile of UML as described in Section A.6.

A.6 ARCHITECTURE OF THE LANGUAGE

SysML is treated as a full modeling language. Formally, it is only a profile of UML. The stereotypes provide extensions to UML elements to address specific needs for system modeling. For instance, UML does not have a model element for requirements or system blocks. Since UML is basically for the modeling of software systems, it provides many elements that are not suitable for system modeling. Therefore, SysML reuses only a subset of the UML. In summary, SysML is a smaller modeling language than UML (Figure A.30).

Figure A.31 shows the relationship of SysML to related packages of the modeling landscape. The SysML profile imports the UML2 metamodel and the standard profile. The standard profile is a set of stereotypes that are specified in the UML specification and are commonly used. The SysML stereotypes extend UML2 metamodel elements or specializes stereotypes from the standard profile. For instance, the SysML trace relationship is a specialization of the trace stereotype from the standard profile (Figure A.29).

SysML also defines some model libraries. The "QUDV" model library provides elements for quantities, units, dimensions, and values. The "ISO 80000" library provides basic units from the corresponding ISO standard.

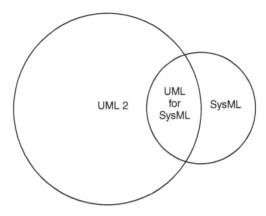

Figure A.30. UML for SysML.

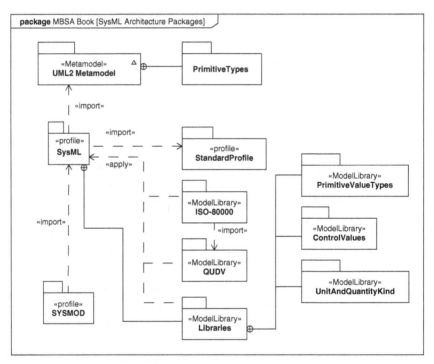

Figure A.31. SysML and UML.

The library "PrimitiveValueTypes" contains the values types "Real," "Integer," "Complex," "String," and "Boolean." The library "ControlValues" contains special elements for activity modeling and the model library "UnitAndQuantityKind" defines the types to define units and quantity kinds.

Appendix **B**

The V-Model

Today, the Systems Engineering Vee is omnipresent in almost every systems development environment. It has an iconic status. The V-Model, as the Systems Engineering Vee is also called, emphasizes a rather natural problem-solving approach. Starting on coarse grain level partitioning the problem to manageable chunks. From the fine grain level the final solution integrates up to the initial level. On each level, one can compare the solution or part of it with the problem or the related part of it. The simplicity of this model (see Figure B.1) permits various projections and results in many interpretations.

B.1 A BRIEF HISTORY OF THE V-MODEL OR THE SYSTEMS ENGINEERING VEE

The V-Model emerged probably in the 1960s, though there seem to be no public citations available. The citations hereafter suggest that the V-Model emerged from more than one source independently. The designation of the model varies depending on the sources. Hereafter, we cite the designation as used in the referenced documents.

Model-Based System Architecture, First Edition.
Tim Weilkiens, Jesko G. Lamm, Stephan Roth, and Markus Walker.
© 2016 John Wiley & Sons, Inc. Published 2016 by John Wiley & Sons, Inc.

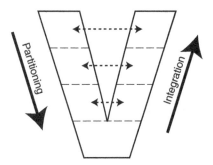

Figure B.1. A basic V-Model.

In 1979, Barry W. Boehm published a paper [15] that was built up on the Vee. He used the Vee in the context of software engineering to emphasize the importance of verification and validation. Boehm made a distinction between an upper part of the Vee for validation and a lower part of the Vee for verification and linked these processes to the related requirements and specifications, respectively. The multilevel nature of systems depicted in the Vee is not further elaborated. Boehm attributed the "V-Chart", as he named it, to personal communication from J.B. Munson, System Development Corporation in 1977.

In a systems context, the "V-Chart" was presented at the first annual conference of NCOSE in 1991 [40]. This conference was the predecessor of today's INCOSE International Symposium. Kevin Forsberg and Harold Mooz introduced the "V-Chart" to clarify the role and responsibility of system and design engineering within the cycle of a project. Other than Boehm's paper that focused on V&V processes in software engineering, Forsberg and Mooz focused on projects realizing systems. Consequently, the multilevel nature of systems is a major issue within this paper. They presented a three-dimensional "Vee-Model" intending to explain the relation between project management and engineering. It acknowledges the iterative and incremental nature of engineering and hence encourages concurrent engineering. Forsberg and Mooz mention a major contribution to the "V-Chart" by Richard Roy.

In 1992, Germany published a standard for governmental IT projects. Apparently, the standard originated from the Ministry of Defense and was there established in the very early 1990s or even earlier. This comprehensive standard is a process description designated as "V-Modell®". The Vee does not stand for the graphical representation of the process model. "V-Modell®" is an abbreviation for the German

"Vorgehensmodell" what could be translated as "process model." Nevertheless, the described processes are depicted in a V-shape. The 1992 edition covered only SW development. It was made available to the public by Bröhl and Dröschel in "Das V-Modell" [18] in 1993. A further developed edition of the "V-Modell®", published in 1997, included systems aspects with references to the system definition by ISO 12207. The standard evolved and became the "V-Modell® XT" with the XT standing for extendable or extreme tailoring. The 1997 edition of the "V-Modell®" as well as the current edition of the "V-Modell® XT" are accessible on a dedicated web site [27], in German and partly in English.

At the INCOSE International Symposium 2013, Dieter Scheithauer and Kevin Forsberg presented the paper "V-Model Views" [124]. This paper collects experiences and improvements from the preceding two decades. The authors extend the scope of the V-Model from the development process to the life cycle of the system, reaching from stakeholders' needs to stakeholders' satisfaction. Compared to the paper of 1991, it splits the overall view into four distinct views: Basic-V, Development-V, Assurance-V, and Dynamic-V. It transforms the horizontal dimension from a time or maturity sequence to a logical sequence of the value stream. Scheithauer and Forsberg emphasize the inductive design of left side of the Vee in opposition to a deductive decomposition imposed by a waterfall-like process. And finally, it exemplifies that validation does not only apply for the finally deployed system but also to each artifact along the life cycle of the system. That is, stakeholder requirements, system requirements, architecture on each level, and integrated into the whole system architecture as well as the operating system need to be validated, meaning checked for their fitness for purpose. The authors state in the paper that the V-Model had been introduced twice, in the 1980's by NASA and with paper by Forsberg and Mooz in 1991.

B.2 A HANDY ILLUSTRATION BUT NO COMPREHENSIVE PROCESS DESCRIPTION

The Systems Engineering Vee is an illustration that depicts only some aspects of the development process of a system. Apart from the German "V-Modell®", the Systems Engineering Vee is no comprehensive process model or process description. Probably the most important aspect depicted in the Vee is the multilevel nature of systems. Based on the definition, a system needs to comprise at least two levels. Therefore, the simplest Vee considering the system only would comprise two levels, the system level and the system element level.

The levels of the Vee correlate with the system levels. This may include also logical levels. The designation of these levels varies depending on the context of application. But each consecutive level pair represents a system in its own right. The lowest level is a bit special, as parts or components represented by this level will not further be partitioned from the viewpoint of the depicted Vee. This imposes that elements represented by the lowest level can be acquired, produced, or harvested. The composition or aggregation of these elements does not matter for the development system of interest. This does not preclude the lowest level elements to be also systems viewed from a different viewpoint. As with each model, the extent of a V-Model should follow a purpose. It typically includes the levels for which a certain team or organization is responsible plus the adjacent lower level and sometimes the adjacent upper level.

Depicting development-related life cycle processes of systems within the Vee, each level needs to comprise instances of the same processes. On each level, requirements engineering, architecting, integration, and verification needs to be performed. The concrete naming of these processes will vary depending on the organization applying these processes. As a consequence, artifacts resulting from these processes will exhibit the Zig-Zag pattern as described in Section 7.1. Figure B.2 depicts a Vee with the basic development processes creating artifacts as named in Figure 7.4. It displays that the Zig-Zag pattern not only exists on the left side of the Vee but proceeds up the right side as well as the system being integrated.

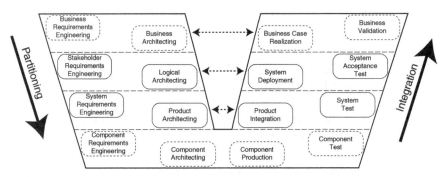

Figure B.2. V-Model with exemplary named basic development processes.

The dotted lines in Figure B.2 designate processes not in the responsibility of the assumed development. The system levels addressed in the Vee are from the top:

- The business as a system to earn money
- A stakeholder process gaining benefit from using the product to be developed
- The product to be developed
- Components from which the product is built.

Figure B.2 does not serve as a development process description. Many important elements are elided. It does not depict any control or object flows. It only maps processes to system levels. The figure designates a requirements engineering process but does not make any distinction between stakeholder and system requirements. In the absence of any object flows it makes no distinction between allocated requirements and elicited requirements. The integration processes are shown with a variety of names and verification and validation processes are mixed and partly designated as test. Figure B.2 follows a widely used, but rather suboptimal naming. Test is used as a synonym for verification and validation, imposing verification or validation are only effective by testing. This neglects other, often more efficient and less costly methods such as inspection, analysis, or demonstration. Finally, validation is reduced to an activity at the very end, though it should start during the stakeholder requirements definition process. Validation happens virtually always within other processes, though sometimes as an isolated task.

The rather simple illustration depicted above neglects a further issue. The number of elements in a lower level is higher than in levels above. Reusing elements in modular systems will degrade the increase of element kinds in subsequent levels. Each of these elements has its own life cycle. Their development or integration sequence depends on availability of data from the adjacent elements. For a comprehensive illustration representing the multiplicity of element kinds, a three-dimensional Vee need to be drawn. This third dimension would consider the interaction between the processes related to different elements on the same level. One can easily imagine, that an illustration with a three-dimensional Vee as depicted in Figure B.3 enriched with processes as displayed above becomes difficult to read and understand. Considering the elided items mentioned before would make it even worse.

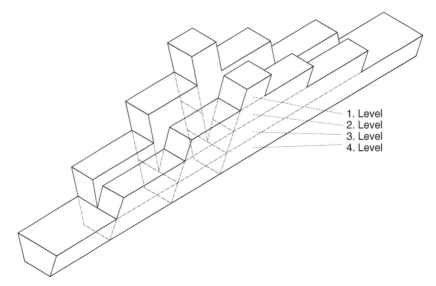

Figure B.3. V-Model considering discrete levels and numbers of elements.

B.3 CRITICAL CONSIDERATIONS

The V-Model received many interpretations. Many of them did not consider the original purpose and criticized the model for not fitting another purpose. The world evolved during the decades since the introduction of the V-Model. New or refined engineering methods came up, new technology and tools provide possibilities to improve effectivity and efficiency of our work. The V-Model can still help to illustrate the multilevel nature of systems and dependencies during the development process. But this can hardly be illustrated in a single view. Dieter Scheithauer and Kevin Forsberg provided with their paper "V-Model Views" [124] a very good explanation and summary for an appropriate application of the V-Model. We address hereafter some issues that frequently lead to discussions.

B.3.1 The V-Model as Process Description

The Systems Engineering Vee, unlike the German "V-Modell®", was never intended as a comprehensive process description. It is intended to depict the multiple levels of a system and only some specific process aspects related to such levels and their dependencies. The V-Model can assist in explaining that the life cycle processes need to be applied at each level for each element belonging to these levels.

B.3.2 The V-Model does not Impose a Waterfall Process

Interpreting the V-Model as equal or similar as a waterfall approach, neglects the meaning of levels within the V-Model. In a waterfall approach, each process, requirements engineering, architecting, implementation, verification, and final validation is considered on its own level. A waterfall process is intended to flow from the top (requirements engineering) down to the bottom (final validation). When comparing such approach with the V-Model, the second half of that cascade is bended up to shape a Vee. This brings requirements engineering and final validation to the top level, followed by architecting and verification on the intermediate level and implementation at the tip of the Vee. This view neglects the meaning of the levels to represent the levels of the system rather than some artificial hierarchy of processes. The horizontal dimension of the Vee represents a logical sequence in the value stream and not a time line of the development process. The inevitably required communication when allocating requirements to lower levels and subsequently validating them requires and ensures a bidirectional exchange of data. Considering the life cycle of each level and hence applying recursively the life cycle processes at each level results in a distinction between allocated requirements and other stakeholder requirements. The upper level becomes one and only one among other stakeholders of the lower level. Unlike with a waterfall approach, where all requirements are supposed to be available up front, the V-Model permits and requires induction of stakeholder requirements at lower levels.

B.3.3 The V-Model Accommodates Iterations

Maintaining a consistent configuration baseline on each level enables iterations in several forms. Iterations are possible on system element level on either side of the Vee. They can involve one or more levels. But also the big iterations including both sides over one or more levels are possible. Especially, the small iteration loops build up on early validation, such as checking requirements allocation and design of system elements for their fitness to contribute to the overall goal. Virtual integration of such system elements in the system model permits validation of each system level and each system element before their implementation starts. Iterative approach in the development of systems is not new and had been addressed already by the invention of Scrum [133]. Forsberg and Mooz promoted iterative development in their 1991-paper [40]. The same applies for early validation of artifacts. Though not described in

detail, Boehm mentioned in his 1979-paper [15] as example requirements validation, design validation, and validation tests for the final
software. He did already emphasize the benefits of early validation.

B.3.4 The V-Model Permits Incremental Development

Using system models and applying iterations as explained above
permits incremental approaches. The use of system models permits
simulations and demonstration of the virtual integrated system. This
permits early validation of the system with the related stakeholders.
Incremental development was already promoted by Forsberg and
Mooz in 1991 [40].

B.3.5 The V-Model and Concurrent Engineering

Since the V-Model emphasizes system levels, each with a number of system elements with their life cycle, it can be used to explain the impact
of concurrent engineering. Maintaining a consistent configuration
baseline and defining increments for each iteration enables concurrent
engineering.

B.3.6 The V-Model Accommodates Change

An incremental and iterative approach and the maintenance of a
consistent configuration baseline permits to predict impact of injected
changes. The V-Model can assist in explaining where change can be
expected. And it can visualize where these changes impact the development. Sources of change include discovering of new stakeholders,
changing stakeholder needs, each validation or verification step.

B.3.7 The V-Model Permits Early Verification Planning

An incremental and iterative approach and the maintenance of a consistent configuration baseline permits early verification planning. Consequently, optimization of the verification processes and related infrastructure can be achieved. This includes combining verifications of different levels or combining interface verification during integration with
system or system element verification.

B.3.8 The V-Model Shows where to Prevent Defects

The V-Model can display where validation should be performed.
Validation does not only apply to the top-rightmost part of the Systems
Engineering Vee. Each artifact created during a development should

be validated. That is, evidence should be provided that the considered artifact contributes to satisfy stakeholders. Early validation ensures that the right problem is solved. Literature does often not very explicitly mention such early validation. The standard ISO/IEC 15288:2008 [60] mentions a number of artifacts to be validated, though only in notes and not in the normative text. The INCOSE Systems Engineering Handbook [56] designates such early validation as "in-process validation" or "continuous validation." The standard ISO/IEC/IEEE 29148:2011 [63] emphasizes the necessity of requirements validation. Scheithauer and Forsberg make the execution of validation very explicit in their Assurance-V [124]. Validation, and hence defect prevention starts at the top-leftmost part of the V-Model and proceeds all the way down and up again, even into the operation phase of the system of interest. Six kinds of validations can be identified along a system life cycle:

1. Validation of stakeholder requirements
2. Validation of allocated requirements
3. Validation of system element definitions
4. Validation of the virtually integrated system
5. Validation of the operational system deployed into its environment
6. Validation of the in-service system.

B.4   READING INSTRUCTION FOR A MODERN SYSTEMS ENGINEERING VEE

To summarize this chapter, we provide a reading instruction for V-Models. As mentioned earlier, a single view can hardly depict each aspect. Views have to be designed to frame concern of a dedicated audience and elide other aspects. Following the seven rules stated hereafter will support a consistent understanding.

B.4.1   The Vertical Dimension

The vertical dimension of the V-Model denotes the multiple levels of the system of interest. The topmost level represents the system context in which the system operates. The bottommost level represents the parts that can be obtained and hence suffice to be defined in a black-box-view. The levels in between represent physical or logical levels of the system. Therefore, the V-Model should comprise of at least three levels: system context, system, and system elements.

B.4.2 The Horizontal Dimension

The horizontal dimension of the V-Model denotes a logical sequence of the value stream. This does not impose that each requirement needs to be frozen upfront. An increment in development may impose to develop some parts before defining the remaining requirements on top level.

B.4.3 The Left Side

The left side of the V-Model denotes the general direction of the top-down development. This is only the general direction. Especially, interface-related issue require to push parts of the development to very low levels before proceeding with the remaining parts of the upper levels.

B.4.4 The Right Side

The right side of the V-Model denotes the general direction of the bottom-up integration. A model-based approach enables virtual integration. This results in a number of instances of the right side. Such representations are sometimes called Y-Model as such virtual integration can be depicted with a branch starting in the middle of the left side and heading upward in parallel to the right side of the V.

B.4.5 The Levels

The Levels represent the system levels. Life cycle processes are applied on each level. Levels get requirements allocated from the next level above, elicit requirements from its specific stakeholders and allocate requirements to the next level below. Levels receive verified system elements (or parts of) from the next level below and provide the integrated and verified system (or parts of) to the next level above.

B.4.6 Life Cycle Processes

The life cycle processes such as stakeholder requirements definition, requirements analysis, architectural design, integration, verification, validation, and the related artifacts appear on each level.

B.4.7 The Third Dimension

The third dimension of the V-Model can be used for different purposes. For instance to display the multitude of system elements per level resulting in parallel application of the life cycle processes.

Bibliography

(1) J. Abulawi. Personal communication. February 12, 2014.

(2) R. L. Ackoff. *Creating the Corporate Future: Plan or be Planned For,* Copyright (C) 1981 by John Wiley & Sons, Inc.

(3) S. Aeschbacher, S. Eisenring, D. Endler, M. Frikart, N. Krüger, J. G. Lamm, and M. Walker. Eine einfache Vorlage zur Architekturdokumentation. In M. Maurer and S.-O. Schulze, editors, *Tag des Systems Engineering, Bremen, 12-14 November 2014,* pages 85–92, München, Germany, 2014. Carl Hanser Verlag.

(4) C. Alexander. *Notes on the synthesis of Form.* Harvard University Press, Cambridge, MA, 1971 (paperback reprint of the 1964 edition).

(5) C. Alexander. *A Pattern Language.* Oxford University Press, New York, 1977.

(6) T. J. Allen. *Managing the Flow of Technology: Technology Transfer and the Dissemination of Technological Information Within the R&D Organization.* Massachusetts Institute of Technology, 1st edition, 1977. Now in a revised edition, 1984 from MIT Press.

(7) G. Altshuller. *Innovation Algorithm: TRIZ, Systematic Innovation and Technical Creativity.* Technical Innovation Center, 1999.

(8) S. W. Ambler. *Agile modeling.* John Wiley & Sons, Inc., New York, 2002.

(9) M. M. Andreasen. *Syntesemetoder på systemgrundlag (Machine Design Methods Based on a Systematic Approach)*. PhD thesis, Lund Technical University, Sweden, 1980.

(10) L. Bass, P. Clements, and R. Kazman. *Software Architecture in Practice*. Addison-Wesley Professional, 3rd edition, 2012.

(11) E. N. Baylin. *Functional Modeling of Systems*. Gordon and Breach, New York, 1990.

(12) R. Beasley, I. Cardow, M. Hartley, and A. Pickard. Structuring requirements in standard templates. In J. Lalk, editor, *EMEASEC 2014 27 – 30 October 2014, Systems Engineering – Exploring New Horizons*. INCOSE, 2014.

(13) B. S. Blanchard. *System Engineering Management*. John Wiley & Sons, Inc., New York, 2004.

(14) H. Blume, H. T. Feldkaemper, and T. G. Noll. Model-based exploration of the design space for heterogeneous systems on chip. *Journal of VLSI Signal Processing*, 40:19–34, 2005.

(15) B. W. Boehm. Guidelines for verifying and validating software requirements and design specification. In *Proceedings of the European Conference on Applied Information Technology of the International Federation for Information Processing (Euro IFIP)*, volume 1, pages 711–719, 1979.

(16) B. W. Boehm. A spiral model of software development and enhancement. *Computer*, 21(5):61–72, 1988.

(17) Tate Britain (Tate Gallery). Press Release – Jimmy Wales announces robots project as the winner of first IK Prize. http://www.tate.org.uk/about/press-office/press-releases/jimmy-wales-announces-robots-project-winner-first-ik-prize, published February 7th, 2014, accessed 14 August 2014.

(18) A.-P. Bröhl and W. Dröschel, editors. *Das V-Modell*. R. Oldenbourg Verlag, München, Germany / Wien, Austria, 1993.

(19) D. M. Buede. *The Engineering Design of Systems*. John Wiley & Sons, Inc., New York, 2009.

(20) P. Clements, R. Kazman, and M. Klein. *Evaluating Software Architectures: Methods and Case Studies*. Addison-Wesley Professional, 3rd edition, 2001.

(21) R. J. Cloutier and D. Verma. Applying the concept of patterns to systems architecture. *Systems Engineering*, 10(2):138–154, 2007.

(22) A. Cockburn. *Crystal Clear: A Human-Powered Methodology for Small Teams*. Addison-Wesley, 2005.

(23) J. M. Cohen and M. J. Cohen. *The Penguin Dictionary of Twentieth-Century Quotations*. VIKING / Penguin Books, 1993.

(24) M. Conrad. Systematic testing of embedded automotive software - the classification-tree method for embedded systems (CTM/ES). In

E. Brinksma, W. Grieskamp, and J. Tretmans, editors, *Perspectives of Model-Based Testing*, Number 04371 in Dagstuhl Seminar Proceedings, Dagstuhl, Germany, 2005, Internationales Begegnungs- und Forschungszentrum für Informatik (IBFI), Schloss Dagstuhl, Germany.

(25) L. L. Constantine. Segmentation and design strategies for modular programming. In T. O. Barnett and L. L. Constantine, editors, *Modular Programming: Proceedings of a National Symposium*. Information & Systems Press, 1968.

(26) J. Daniels and T. Bahill. The hybrid process that combines traditional requirements and use cases. *Systems Engineering*, 7(4):303–319, 2004.

(27) Das V-Modell®. http://v-modell.iabg.de/, accessed 8 November 2014.

(28) J. Dick and J. Chard. The systems engineering sandwich: combining requirements, models, and design. In *Proceedings of the fourteenth annual International Symposium of the International Council on Systems Engineering*, 2004.

(29) C. E. Dickerson and D. N. Mavris. *Architecture and Principles of Systems Engineering, Complex and Enterprise Systems Engineering*. Auerbach Publications Taylor & Francis Group, 2010.

(30) M. Dänzer, W. Gerritsen, J. G. Lamm, and T. Weilkiens. Funktionale Architektur trifft Schichtenarchitektur. In M. Maurer and S.-O. Schulze, editors, *Tag des Systems Engineering, Paderborn, 7-9 November 2012*, pages 363–372, München, Germany, 2012. Carl Hanser Verlag.

(31) M. Dänzer, S. Kleiner, J. G. Lamm, G. Moeser, F. Morant, F. Munker, and T. Weilkiens. Funktionale Systemmodellierung nach der FAS-Methode: Auswertung von vier Industrieprojekten. In M. Maurer and S.-O. Schulze, editors, *Tag des Systems Engineering, Bremen, 12-14 November 2014*, pages 109–118, München, Germany, 2014. Carl Hanser Verlag.

(32) S. Eisenring, M. Frikart, W. Gerritsen, C. Krainer, J. G. Lamm, A. Mettauer, M. Walker, and M. Zollinger. Auf dem Weg zu einem Leitfaden im Systems Engineering für moderat-komplexe Systeme. In M. Maurer and S.-O. Schulze, editors, *Tag des Systems Engineering, Paderborn, 7-9 November 2012*, pages 33–42, München, Germany, 2012. Carl Hanser Verlag.

(33) H. Eisner. *Managing Complex Systems: Thinking Outside the Box*. John Wiley & Sons, Inc., Hoboken, NJ, 2005.

(34) J. P. Elm and D. R. Goldenson. The business case for systems engineering study: results of the systems engineering effectiveness survey. Technical Report CMU/SEI-2012-SR-009, Software Engineering Institute, 2012.

(35) M. R. Emes. Strategic multi-stakeholder trade studies. In *Proceedings of EuSEC*, 2006.

(36) M. R. Emes, P. A. Bryant, M. K. Wilkinson, P. King, A. M. James, and S. Arnold. Interpreting "Systems Architecting". *Systems Engineering*, 15(4):369–395, 2012.

(37) Federation of EA Professional Organizations (FEAPO). Common perspectives on enterprise architecture. *Architecture and Governance Magazine*, (9-4), 2013.

(38) J. L. Fernández-Sánchez, M. García-García, J. García-Muñoz, and J. P. Gómez-Pérez. La ingeniería de sistemas y su aplicación a un vehículo aéreo no tripulado. *Dyna*, 87(4):456–466, 2012.

(39) D. G. Firesmith, P. Capell, D. Falkenthal, C. B. Hammons, D. T. Latimer IV, and T. Merendino. *The Method Framework for Engineering System Architectures*. CRC Press, 2009.

(40) K. Forsberg and H. Mooz. The relationship of systems engineering to the project cycle. In *Proceedings of the first annual conference of NCOSE*, 1991.

(41) K. Forsberg, H. Mooz, and H. Cotterman. *Visualizing Project Management*. John Wiley & Sons, Inc., New York, 2005.

(42) J. Fried and D. H. Hansson. *REWORK*. Crown Business, 2010.

(43) D. Garlan and M. Shaw. An introduction to software architecture. Technical Report CMU/SEI-94-TR-21, Carnegie Mellon University, 1994.

(44) D. Gianni, N. Lindman, J. Fuchs, and R. Suzic. Introducing the European space agency architectural framework for space-based systems of systems engineering. In O. Hammami, D. Krob, and J.-L. Voirin, editors, *Complex Systems Design & Management*, pages 335–346. Springer-Verlag, 2012, Berlin and Heidelberg.

(45) J. L. Gibson, J. M. Ivancevich, J. H. Donnelly, and R. Konopaske. *Organizations: Behaviour, Structure, Processes*. McGraw-Hill, 14th edition, 2012.

(46) GOV.UK. MOD Architecture Framework - Detailed guidance, 2012, https://www.gov.uk/mod-architecture-framework, accessed 20 February 2015.

(47) D. Gray and T. V. Wal. *The Connected Company*. O'Reilly, 2012.

(48) D. Greefhorst and E. Proper. *Architecture Principles – The Cornerstones of Enterprise Architecture, The Enterprise Engineering Series*. Springer-Verlag, Berlin and Heidelberg, 2011.

(49) M. Grundel, J. Abulawi, G. Moeser, T. Weilkiens, A. Scheithauer, S. Kleiner, C. Kramer, M. Neubert, S. Kümpel, and A. Albers. FAS4M – no more: "Please mind the gap!". In M. Maurer and S.-O. Schulze, editors, *Tag des Systems Engineering, Bremen, 12-14 November 2014*, pages 65–74, München, Germany, 2014. Carl Hanser Verlag.

(50) C. Hampden-Turner and F. Trompenaars. Response to Geert Hofstede. *International Journal of Intercultural Relations*, 21(1):149–159, 1997.

(51) M. Hassenzahl, A. Beu, and M. Burmester. Engineering joy. *IEEE Software*, 18(1):70–76, 2001.

(52) L. Herrero. *Viral Change^TM The Alternative to Slow, Painful and Unsuccessful Management of Change in Organisations.* meetingminds, 2006, 2008.

(53) D. K. Hitchins. *Systems Engineering.* John Wiley & Sons, 2007.

(54) INCOSE MBSE Challenge Team SE2. *Cookbook for MBSE with SysML*, 2011.

(55) INCOSE. Systems engineering vision 2020. Available at http://www .incose.org/ProductsPubs?products/sevision2020.aspx, September 2007. INCOSE-TP-2004-004-02, accessed 1st June 2015.

(56) INCOSE. *INCOSE Systems Engineering Handbook v. 3.2*, January 2010.

(57) INCOSE. Systems engineering vision 2025: the world in motion, June 2014. Available at http://www.incose.org/newsevents/announcements /docs/SystemsEngineeringVision_2025 _June2014.pdf, accessed 1st June 2015.

(58) ISO. ISO Online Browsing Platform (OBP). http://www.iso.org/obp, accessed 4 January 2014.

(59) ISO 13849-1:2006. Safety of machinery — Safety-related parts of control systems — Part 1: General principles for design, 2006.

(60) ISO/IEC 15288:2008 and IEEE Std 15288-2008. Systems and software engineering — System life cycle processes, 2008.

(61) ISO/IEC TR 24748-1:2010. Systems and software engineering — life cycle management — Part 1: Guide for life cycle management, 2010.

(62) ISO/IEC 7498-1:1984. Information processing systems — Open Systems Interconnection — Basic Reference Model, 1984.

(63) ISO/IEC/IEEE 29148:2011. Systems and software engineering — Life cycle processes — Requirements engineering, 2011.

(64) ISO/IEC/IEEE 42010:2011. Systems and software engineering — Architecture description, 2011.

(65) IEEE Std 1471-2000. IEEE Recommended Practice for Architectural Description of Software-Intensive Systems, 2000.

(66) D. D. Jackson, P. Watzlawick, and J. B. Bavelas. *Pragmatics of Human Communication: A Study of Interactional Patterns, Pathologies and Paradoxes.* W. W. Norton, 2011.

(67) I. Jacobson, M. Christerson, P. Jonsson, and G. Övergaard. *Object-Oriented Software Engineering - A Use Case Driven Approach.* Addison-Wesley, 1992.

(68) C. G. Jung. *Psychologische Typen.* Rascher Verlag, 1921.

(69) C. G. Jung. *Psychological Types, Bollingen Series,* vol. 20. Pantheon Books, 1971.

(70) K. C. Kang, S. G. Cohen, J. A. Hess, W. E. Novak, and A. S. Peterson. Feature-oriented domain analysis (FODA) feasibility study. Technical Report CMU/SEI-90-TR-21, Software Engineering Institute, 1990.

(71) K. C. Kang, J. Lee, and P. Donohoe. Feature-oriented product line engineering. *IEEE Software,* 19:58–65, 2002.

(72) R. Kazman, M. Klein, and P. Clements. ATAM: method for architecture evaluation. Technical Report CMU/SEI-2000-TR-004, Carnegie Mellon University, 2000.

(73) T. Kelley. *The Ten Faces of Innovation.* Profile Books, 2006.

(74) M. Kennedy. *Tate Offers Chance to Experience Night at the Museum, with the Help of Four Robots.* The Guardian (London / Manchester edition, United Kingdom), August 13, 2014.

(75) K. Keutzer, S. Malik, A. R. Newton, J. M. Rabaey, and A. Sangiovanni-Vincentelli. System-level design: orthogonalization of concerns and platform-based design. *IEEE Transactions on Computer-Aided Design of Integrated Circuits and Systems,* 19(12):1523–1543, 2000.

(76) H. Kissinger. *Diplomacy.* Touchstone / Simon & Schuster, 1994.

(77) D. A. Kolb. *Experiential Learning: Experience as the Source Of Learning And Development.* Prentice Hall, Englewood Cliffs, NJ, 1984.

(78) A. Korff, J. G. Lamm, and T. Weilkiens. Werkzeuge für den Schmied funktionaler Architekturen. In M. Maurer and S.-O. Schulze, editors, *Tag des Systems Engineering, Hamburg, 9-11 November 2011,* pages 3–12, München, Germany, 2011. Carl Hanser Verlag.

(79) A. Kossiakoff and W. N. Sweet. *Systems Engineering — Principles and Practice.* John Wiley & Sons, Inc., Hoboken, NJ, 2003.

(80) R. Krikhaar, W. Mosterman, N. Veerman, and C. Verhoef. Enabling system evolution through configuration management on the hardware/software boundary. *Systems Engineering,* 12(3):233–264, 2009.

(81) P. B. Kruchten. The 4+1 view model of architecture. *IEEE Software,* 12(6):42–50, 1995.

(82) J. G. Lamm, A. Lohberg, and T. Weilkiens. Funktionale Architekturen in der Systementwicklung anwenden. In M. Maurer and S.-O. Schulze, editors, *Tag des Systems Engineering, Stuttgart, 6-8 November 2013,* pages 283–292, München, Germany, 2013. Carl Hanser Verlag.

(83) J. G. Lamm and T. Weilkiens. Funktionale Architekturen in SysML. In M. Maurer and S.-O. Schulze, editors, *Tag des Systems Engineering, München Freising, 10-12 November 2010,* pages 109–118, München, Germany, 2010. Carl Hanser Verlag.

(84) J. G. Lamm and T. Weilkiens. Method for deriving functional architectures from use cases. *Systems Engineering*, 17(2):225–236, 2014.

(85) C. Larman and B. Vodde. *Practices for Scaling Lean and Agile Development*. Addison-Wesley / Pearson, 2010.

(86) H. 'Bud' Lawson. *A Journey Through the Systems Landscape, Systems Thinking and Systems Engineering*, vol. 1. 2010.

(87) G. H. Lewes. *Problems of Life and Mind*. Number 2 in 1. Mifflin/Trübner, Houghton, MI, 1875.

(88) V.-C. Liang and C. J. J. Paredis. A port ontology for conceptual design of systems. *Journal of Computing and Information Science in Engineering*, 4:206–217, 2004.

(89) T. Liland, H. D. Jørgensen, and S. Skogvold. Aligning TOGAF and NAF - experiences from the Norwegian Armed Forces. In P. Johannesson, J. Krogstie, and A. L. Opdahl, editors, *The Practice of Enterprise Modeling, Lecture Notes in Business Information Processing*, Vol. 92, pages 131–146. Springer-Verlag, 2011, Berlin and Heidelberg.

(90) J. Lockett and J. Powers. Human factors engineering methods and tools. *Handbook of Human Systems Integration*, pages 463–496. John Wiley & Sons, Inc., 2003.

(91) M. W. Maier. System and software architecture reconciliation. *Systems Engineering*, 9(2):146–159, 2006.

(92) M. W. Maier and E. Rechtin. *The Art of Systems Architecting*. CRC Press, Boca Raton, FL, 2nd edition, 2002.

(93) J. N. Martin. Using the PICARD theory of systems to facilitate better systems thinking. *INCOSE INSIGHT*, 11(1):37–41, 2008.

(94) MBSE Wiki Article – Model Based Document Generation, 2013, http://www.omgwiki.org/MBSE/doku.php?id=mbse:telescope#model_based_document_generation, accessed 19 June 2014.

(95) B. McCarthy. About Learning - Official Site of Bernice McCarthy's 4MAT System, 2015. http://www.aboutlearning.com, accessed 15 February 2015.

(96) G. A. Miller. The magical number seven, plus or minus two: some limits on our capacity for processing information. *Psychological Review*, 63(2):81–97, 1956.

(97) J. P. Monat. Why customers buy − a look at how industrial customers make purchase decisions. *Marketing Research*, 28:20–24, 2009.

(98) G. Muller. *Systems Architecting − A Business Perspective*. CRC Press, 2011.

(99) *National Airspace System - System Engineering Manual (SEM), v3.1, 2006.*

(100) NATO Architecture Framework v4.0 Documentation. http://nafdocs.org, accessed 20 February 2015.

(101) Object Management Group (OMG). Business Motivation Model (BMM) Version 1.2. OMG Document Number formal/2014-05-01, 2014.

(102) Object Management Group (OMG). Common variability language (CVL) revised submission. OMG document number ad/2012-08-05, 2012.

(103) Object Management Group (OMG). MDA Guide Version 1.0.1. OMG Document Number omg/2003-06-01, 2003.

(104) Object Management Group (OMG). Meta Object Facility (MOF) Version 2.4.2. OMG Document Number formal/2014-04-03, 2014.

(105) Object Management Group (OMG). OMG Systems Modeling Language (OMG SysML) Version 1.4 - Beta. OMG Document Number ptc/2013-12-09, 2013.

(106) Object Management Group (OMG). OMG Unified Modeling Language (OMG UML) Version 2.4.1. OMG Document Number formal/2011-08-13, 2011.

(107) Object Management Group (OMG). Requirements Interchange Format (ReqIF) Version 1.1. OMG Document Number formal/2013-10-01, 2013.

(108) Object Management Group (OMG). Semantics of a Foundational Subset for Executable UML Models (FUML) Version 1.1, OMG Document Number formal/2013-08-06, 2013.

(109) Object Management Group (OMG). XML Metadata Interchange (XMI) Version 2.4.2. OMG Document Number formal/2014-04-04, 2014.

(110) B. W. Oppenheim. *Lean for Systems Engineering with Lean Enablers for Systems Engineering*. John Wiley & Sons, Inc., 2011.

(111) T. J. Ostrand and M. J. Balcer. The category-partition method for specifying and generating functional tests. *Communications of the ACM*, 31(6):676–686, 1988.

(112) Oxford Dictionaries. Oxford Dictionaries – Definition of architect in English. http://www.oxforddictionaries.com/definition/english/architect, accessed 30 July 2014.

(113) G. Pahl, W. Beitz, J. Feldhusen, and K.-H. Grote. *Engineering Design*. Springer-Verlag, 2007.

(114) D. L. Parnas. On the criteria to be used in decomposing systems into modules. *Communications of the ACM*, 15(12):1053–1058, 1972.

(115) G. Patzak. *Systemtechnik, Planung Komplexer Innovativer Systeme*. Springer-Verlag, 1982.

(116) N. Plum. Trak enterprise architecture framework, 2010, http://trak.sourceforge.net, accessed 30 July 2014.

(117) K. Pohl. *Requirements Engineering - Fundamentals, Principles, and Techniques*. Springer-Verlag, 2010.

(118) K. Pohl, G. Böckle, and F. J. Linden. *Software Product Line Engineering: Foundations, Principles and Techniques.* Springer-Verlag, 1st edition, 2005.

(119) V. Pollio. DE ARCHITECTURA LIBER PRIMUS. In *DE ARCHITECTURA LIBRI DECEM*, Chapter 3. most likely first century B.C.

(120) D. J. Richardson and L. A. Clarke. A partition analysis method to increase program reliability. In *Proceedings of the 5th International Conference on Software Engineering*, pages 244–253, San Diego, CA, March 1981. IEEE.

(121) N. Richard, D. George, and E. P. Box. *Empirical Model-Building and Response Surfaces, Wiley Series in Probability and Mathematical Statistics: Applied Probability and Statistics.* John Wiley & Sons, New York, 1987.

(122) S. Robertson and J. Robertson. *Mastering the Requirements Process: Getting Requirements Right.* Addison-Wesley Professional, 3rd edition, 2012.

(123) N. Rozanski and E. Woods. *Software Systems Architecture.* Addison Wesley / Pearson, 2012.

(124) D. Scheithauer and K. Forsberg. V-model views. In *Proceedings of the 23nd Annual International Symposium of the International Council on Systems Engineering*, 2013.

(125) G. Schuh. *Produktkomplexität Managen.* Carl Hanser Verlag München, 2005.

(126) C. E. Shannon. A mathematical theory of communication. *Bell System Technical Journal*, 27, 1948.

(127) D. C. Skinner. *Decision Analysis.* Probabilistic Publishing, 2001.

(128) Software Engineering Institute / Carnegie Mellon University. What is your definition of software architecture? http://www.sei.cmu.edu /architecture/start/glossary/definition-form.cfm, accessed 8 February 2015.

(129) D. E. Spielberg. Methodik zur Konzeptfindung basierend auf technischen Kompetenzen. Doctoral thesis, RWTH Aachen, Shaker-Verlag, 2002.

(130) H. Stachowiak. *Allgemeine Modelltheorie.* Springer-Verlag, 1973.

(131) M. Stickdorn and J. Schneider. *This is Service Design Thinking.* BIS Publishers, Amsterdam, The Netherlands, 2013.

(132) N. P. Suh. *The Principles of Design.* Oxford University Press, New York, 1990.

(133) H. Takeuchi and I. Nonaka. The new new product development game. *Harvard Business Review*, pages 137–146, January / February 1986.

(134) The Myers & Briggs Foundation. 2015, http://www.myersbriggs.org, accessed 15 February 2015.

(135) The White House. Federal Enterprise Architecture (FEA). http://www .whitehouse.gov/omb/e-gov/fea, accessed 17 February 2015.

(136) The Open Group. The Open Group Architecture Framework (TOGAF®) Version 9.1, 2011.

(137) F. Trompenaars and C. Hampden-Turner. *Riding the Waves Of Culture.* Nicholas Brealey Publishing, 2012.

(138) D. G. Ullman. *Making Robust Decisions.* Trafford Publishing, 2006.

(139) K. Ulrich. The role of product architecture in the manufacturing firm. *Research Policy*, 24:419–440, 1995.

(140) UML Profile for Modeling Quality of Service and Fault Tolerance Characteristics and Mechanisms Specification Version 1.2. OMG Document Number formal/2010-06-01, 2010.

(141) U.S. Chief Information Officer. DoDAF - DOD Architecture Framework Version 2.02, 2010, http://dodcio.defense.gov/TodayinCIO /DoDArchitectureFramework.aspx, accessed 20 February 2015.

(142) A. Vollerthun. Design-to-market integrating conceptual design and marketing. *Systems Engineering*, 5(4):315–326, 2002.

(143) F. S. von Thun. *Miteinander reden 1-4: Störungen und Klärungen. Stile, Werte und Persönlichkeitsentwicklung. Das "Innere Team" und situationsgerechte Kommunikation. Fragen und Antworten.* Miteinander reden. Rowohlt Taschenbuch Verlag, 2014.

(144) B. Waber, J. Magnolfi, and G. Lindsay. Der Wert der Gestaltung. *Harvard Business Manager*, 2/2015, 2015.

(145) T. Weilkiens. *Systems Engineering with SysML / UML.* Morgan Kaufmann OMG Press, 2008.

(146) T. Weilkiens. A system architecture is what system architects create (blog entry), 2014, http://model-based-systems-engineering.com/2014 /03/10/a-system-architecture-is-what-system-architects-create/ accessed 3rd January 2015.

(147) T. Weilkiens. *Systems Engineering mit SysML.* dpunkt-Verlag, 2014.

(148) T. Weilkiens. *Variant Modeling with SysML.* Leanpub, 2015.

(149) E. Wenger. *Communities of Practice.* Cambridge University Press, 1998.

(150) A. A. Yassine and L. A. Wissmann. The implications of product architecture on the firm. *Systems Engineering*, 10(2):118–137, 2007.

(151) Zachman International Enterprise Architecture. About the Zachman Framework. https://www.zachman.com/about-the-zachman-framework, accessed 16 February 2015.

(152) C. Zingel, A. Albers, S. Matthiesen, and M. Maletz. Experiences and advancements from one year of explorative application of an integrated model-based development technique using $C\&C^2$-A in SysML. *IAENG International Journal of Computer Science*, 39(2):165–181, 2012.

Index

Model-Based System Architecture, First Edition.
Tim Weilkiens, Jesko G. Lamm, Stephan Roth, and Markus Walker.
© 2016 John Wiley & Sons, Inc. Published 2016 by John Wiley & Sons, Inc.

WILEY SERIES IN SYSTEMS ENGINEERING AND MANAGEMENT

Andrew P. Sage, Editor

ANDREW P. SAGE and JAMES E. ARMSTRONG, Jr.
Introduction to Systems Engineering

WILLIAM B. ROUSE
Essential Challenges of Strategic Management

YEFIM FASSER and DONALD BRETTNER
Management for Quality in High-Technology Enterprises

THOMAS B. SHERIDAN
Humans and Automation: System Design and Research Issues

ALEXANDER KOSSIAKOFF and WILLIAM N. SWEET
Systems Engineering Principles and Practice

HAROLD R. BOOHER
Handbook of Human Systems Integration

JEFFREY T. POLLOCK and RALPH HODGSON
Adaptive Information: Improving Business Through Semantic Interoperability, Grid Computing, and Enterprise Integration

ALAN L. PORTER and SCOTT W. CUNNINGHAM
Tech Mining: Exploiting New Technologies for Competitive Advantage

REX BROWN
Rational Choice and Judgment: Decision Analysis for the Decider

WILLIAM B. ROUSE and KENNETH R. BOFF (editors)
Organizational Simulation

HOWARD EISNER
Managing Complex Systems: Thinking Outside the Box

STEVE BELL
Lean Enterprise Systems: Using IT for Continuous Improvement

J. JERRY KAUFMAN and ROY WOODHEAD
Stimulating Innovation in Products and Services: With Function Analysis and Mapping

WILLIAM B. ROUSE
Enterprise Tranformation: Understanding and Enabling Fundamental Change

JOHN E. GIBSON, WILLIAM T. SCHERER, and WILLAM F. GIBSON
How to Do Systems Analysis

WILLIAM F. CHRISTOPHER
Holistic Management: Managing What Matters for Company Success

WILLIAM B. ROUSE
People and Organizations: Explorations of Human-Centered Design

MOJAMSHIDI
System of Systems Engineering: Innovations for the Twenty-First Century

ANDREW P. SAGE and WILLIAM B. ROUSE
Handbook of Systems Engineering and Management, Second Edition

JOHN A. CLYMER
Simulation-Based Engineering of Complex Systems, Second Edition

KRAG BROTBY
Information Security Governance: A Practical Development and Implementation Approach

JULIAN TALBOT and MILES JAKEMAN
Security Risk Management Body of Knowledge

SCOTT JACKSON
Architecting Resilient Systems: Accident Avoidance and Survival and Recovery from Disruptions

JAMES A. GEORGE and JAMES A RODGER
Smart Data: Enterprise Performance Optimization Strategy

YORAM KOREN
The Global Manufacturing Revolution: Product-Process-Business Integration and Reconfigurable Systems

AVNER ENGEL
Verification, Validation, and Testing of Engineered Systems

WILLIAM B. ROUSE (editor)
The Economics of Human Systems Integration: Valuation of Investments in People's Training and Education, Safety and Health, and Work Productivity

ALEXANDER KOSSIAKOFF,WILLIAM N. SWEET, SAM SEYMOUR, and STEVEN M. BIEMER
Systems Engineering Principles and Practice, Second Edition

GREGORY S. PARNELL, PATRICK J. DRISCOLL, and DALE L HENDERSON (editors)
Decision Making in Systems Engineering and Management, Second Edition

ANDREW P. SAGE and WILLIAM B. ROUSE
Economic Systems Analysis and Assessment: Intensive Systems, Organizations, and Enterprises

BOHOAN W. OPPENHEIM
Lean for Systems Engineering with Lean Enablers for Systems Engineering

LEV M. KLYATIS
Accelerated Reliability and Durability Testing Technology

BJOERN BARTELS , ULRICH ERMEL, MICHAEL PECHT, and PETER SANDBORN
Strategies to the Prediction, Mitigation, and Management of Product Obsolescence

LEVANT YILMAS and TUNCER OREN
Agent-Directed Simulation and Systems Engineering

ELSAYED A. ELSAYED
Reliability Engineering, Second Edition

BEHNAM MALAKOOTI
Operations and Production Systems with Multiple Objectives